T0295223

The Music of Physics

The Music of Physics

An Introduction to the Harmonies of Nature

George N. Gibson

JENNY STANFORD

PUBLISHING

Published by

Jenny Stanford Publishing Pte. Ltd.
101 Thomson Road
#06-01, United Square
Singapore 307591

Email: editorial@jennystanford.com
Web: www.jennystanford.com

British Library Cataloguing-in-Publication Data
A catalogue record for this book is available from the British Library.

The Music of Physics: An Introduction to the Harmonies of Nature

ISBN 978-981-4968-99-7 (Hardcover)
ISBN 978-1-003-48481-3 (eBook)

Dedication

To my wonderfully creative wife and the rest of my family.

Contents

Preface

This textbook originates from the notes of a course taught at the University of Connecticut for over 25 years and more than 5 years as a fully online course. It satisfies the university requirements for science and technology, quantitative reasoning, and laboratory and is intended for non-science majors. The success of the course is manifest by the enthusiastic response of the hundreds of students who have taken the course. The online course has been particularly rewarding for seeing how students, with little experience with science, alone at home, and with only remote assistance, have taken impressive data.

Introduction: Purpose and Scope of the Textbook

This textbook is written for a very specific purpose and audience: it serves as a breadth requirement for a general education program at the college level for non-science majors. It should satisfy requirements for a science and technology course, quantitative reasoning, and optionally a laboratory. It is also suitable for an introductory course at the high school level.

Many traditional physics classes focus on mathematically rigorous, but often unintuitive concepts. For example, calculus is required for a deep and unifying understanding of classical mechanics, but this is not appropriate for an entry-level general education course. Without that deeper level, classical mechanics can leave students with a poor attitude towards physics and science, in general. Given that this course will probably be taken as a requirement, and given that it may be the last science course students take, it is vital to think carefully about what it is trying to achieve, the subject matter, and the academic level. It must be engaging and meaningful to the students and provide a positive experience. Content, for the most part, will not be remembered in detail, while the level of interest and excitement about the course will.

The solution to this problem as proposed by this textbook is to use music as a gateway to various topics in physics and especially modern physics. Interestingly, many aspects of music, and, by extension, waves are understood intuitively and correctly by

The Music of Physics: An Introduction to the Harmonies of Nature
George N. Gibson
Copyright © 2024 Jenny Stanford Publishing Pte. Ltd.
ISBN 978-981-4968-99-7 (Hardcover), 978-1-003-48481-3 (eBook)
www.jennystanford.com

most students. This immediately provides a more welcoming and engaging atmosphere. Even more interesting, music has often, through the centuries, contributed to physics. This breaks down one of the barriers that physics presents to novice scientists and emphasizes that knowledge often comes from an exchange of ideas between different areas of study.

Unlike many introductory texts on the physics of music, this textbook is not restricted to purely musical topics. Rather, each topic is carefully chosen to relate to broader principles applicable to other areas of physics or technology. In this way, there are no intellectual "dead ends". That is, one wants to avoid students asking, "why are we learning this?" about any topic in the course. In terms of content, physics education research has clearly established that less is more. Therefore, this textbook presents a rather restricted number of topics so that students can learn them in some depth. Finally, while knowledge of music is obviously helpful for a course like this, it is strictly not necessary as the textbook covers all necessary musical topics.

Students are often put off by the math in a physics course, and students in such courses tend to have widely varying backgrounds. The math in this textbook is, again, carefully chosen to be accessible to all students. It is the author's viewpoint that derivations of formulas do not usually add to their understanding, even in advanced physics courses. For this reason, there are no derivations in this textbook. However, without derivations, equations often become recipes for substituting in numbers with no appreciation for what they mean. As a result, the Appendix provides practical laboratories to test *every* equation and concept in the course. Although a scientist does not necessarily need to know where an equation came from, he or she needs to be able to test its validity. That is the approach taken here.

Students are still encouraged to use some level of quantitative reasoning throughout this book. One technique is to look at the relationship between variables. For example, given the equation $pV = nRT$, one can ask the following: if I know the initial pressure, what will be the final pressure if the temperature is doubled and everything else is kept constant? This focuses on how pressure depends on temperature and, hence, equations are mostly referred to as dependencies. This also demonstrates a primary goal of the course: how does one proceed in science with incomplete knowledge? Indeed, science means making progress in the face of incomplete knowledge. In the simple example above, one does not need to know the values of V, n, and R or even what these variables mean to find a quantitative answer to the particular question asked.

There are occasional instances where more mathematical knowledge is required. This includes the algebraic manipulation of one equation, a few examples of algebra with two equations, and a very small amount of trigonometry. Units are discussed, as well as generally how to convert units. There are some problems focused on unit conversion, but most of the other problems do not require unit conversion, except centimeters to meters. The point is to keep the math very restricted so as to not detract from the physics.

This textbook is not intended to be an exhaustive or definite treaties on any topic except for the basic physical principles—it is simply a physics course. A famous guiding principle is that clarity and accuracy are inversely related: clarity = 1/accuracy. This course is about clarity. To be more specific: the history of various concepts is provided to stimulate an understanding of how music has influenced physics. However, history is complex and any one topic could fill an entire course. This textbook only paints broad strokes that are specifically relevant to the course. For example, temperament, how keyboards are tuned, is an important concept in the course, as it is an example of how mathematical underpinnings affect areas beyond physics, such as music. However, the topic is, by necessity, rather simplified. For light, distance and optical path length are treated as the same, avoiding a detailed discussion of index of refraction. The goal is to emphasize similarities between different types of waves. However, any time shortcuts are taken, they are identified and briefly discussed.

As mentioned above, this textbook contains an Appendix with an extensive list of laboratories that cover every major concept. Students can perform most of the labs on their own with minimal amounts of material required. For the most part, they use earbuds, a microphone, and free audio software. Some labs can also be performed with a virtual wave simulator available online. While many online labs are very restricted, this particular piece of software is free-form, meaning the students must, essentially, design and run their own experiment. The results are quantitative and demonstrate the power of spectroscopy. Finally, two labs are discussed that require resources appropriate for a school or college laboratory, although one has an excellent virtual counterpart. Moreover, sample data is provided, so the analysis, at least, can be done, even without the physical labs. With the availability of the different kinds of labs, the course can be offered without the labs, with a few labs assigned as homework, or as a full-fledged laboratory course.

In summary, music provides an excellent introduction to physics and the methods of scientific inquiry, as music has been deeply intertwined with physics throughout history. Why music plays such an important role in the development of physics is an intriguing question and this textbook hopefully provides some insight.

Chapter 1

Waves and the Language of Physics

1.1   Pitch, Loudness, Timbre

To begin our discussion, we consider the attributes or characteristics of any kind of wave, although we will use sound waves as an example, as each of these characteristics can be associated with something that we hear in a sound wave.

The most noticeable aspect of a wave is that it repeats in time. Whether it is a vibrating string on a violin or waves breaking at the shore, something is repeating. Each repetition is called an oscillation:

An *oscillation* is one segment of a repetitive motion.

Pitch, Frequency, Period

Musical notes or tones have a *pitch*. Pitch can informally be thought of as how "high" or "low" a musical note sounds, but the pitch of a particular note can be given as a number. For example, the note "A" in the middle of a piano is designated A = 440. Now, the question is 440 what? This number is how many oscillations occur in 1 s.

If you pluck a violin string tuned to Middle A, the string will vibrate or oscillate back and forth 440 times each second, which will produce a certain pitch. The technical term for *pitch* is *frequency* and the frequency refers to *how many times in 1 s the string oscillates back and forth:*

The Music of Physics: An Introduction to the Harmonies of Nature
George N. Gibson
Copyright © 2024 Jenny Stanford Publishing Pte. Ltd.
ISBN 978-981-4968-99-7 (Hardcover), 978-1-003-48481-3 (eBook)
www.jennystanford.com

The pitch of Middle A corresponds to a frequency of 440 oscillations in 1 s.

Physicists do not like to keep write expressions out like this, so a shorthand has been developed. We can write:

A 440 = 440 oscillations per second
= 440 oscillations/second
= 440/s = 440 Hz

Each way of writing this frequency gets progressively more compact. Since frequency always refers to some number of oscillations, we do not have to keep writing "oscillations". Also, "per second" is more easily written as "/second", and second is abbreviated as "s". What may be more unfamiliar is the designation that "/s" = "Hz". Hz is an abbreviation of the unit Hertz, named after the physicist Heinrich Hertz. 1 Hertz corresponds to 1 oscillation/second.

Once we understand the meaning of a pitch or frequency of 440 Hz, we can ask a related question: how long does 1 oscillation of the vibrating string take? If the string oscillates 440 times in 1 s, then each oscillation will take only 1/440 s or approximately 0.0023 s. That's very fast! Another way to look at this is the following: if each oscillation takes (1/440) s then 440 oscillations will take 1s. This is the same as saying that there are 440 oscillations/second.

So, the time it takes for 1 oscillation is called the *period*, and the period is related to the frequency by:

$$\text{Period} = \frac{1}{\text{Frequency}}. \tag{1.1}$$

For the example of a pitch of 440 Hz, the period will be:

$$1/(440/\text{s}) = (1/440)\,\text{s} = 0.00227\,\text{s}$$
$$= 2.27\ \text{millisecond}$$
$$= 2.27\ \text{ms}$$

Again, we have used some shorthand notation. If the period is rather small, we don't want to keep writing lots of zeros after the decimal point, so we use scientific notation, instead: $0.001\,\text{s} = 10^{-3}\,\text{s}$

which corresponds to 1 millisecond and is abbreviated as 1 ms. This is especially convenient for sound waves, as the periods of sound waves are generally around 1 to 100 ms (see Lab 1.1).

Periodic sounds can be formed in different ways. The most common, of course, is from a musical instrument. But, now, consider the situation where you are standing in front of a set of bleachers and you strike a bass drum (Fig. 1.1).

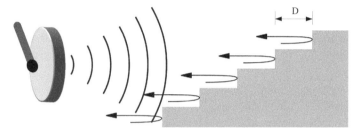

Figure 1.1 Periodic echo of a drum beat reflecting of off a set of steps.

The individual echoes of the drum off each step are delayed from one another. So, to the drummer the echoes form a repetitive sound which then is heard as a pitch. Anyone who has played in a marching band can attest to this.

Amplitude—Loudness

Besides the pitch of a musical note, perhaps the most noticeable feature is how loud the note is. The *loudness* of a sound wave is determined by its *amplitude*. While loudness is only associated with sound waves, all types of waves have an amplitude. For example, water waves on a calm ocean may be less than 1 foot high. Good surfing waves might be 10 feet or more in amplitude. During a storm, the amplitude might increase to 40 or 50 feet. Thus, many things can influence the amplitude. For sound waves, the important considerations are:

1. What is producing the sound?
2. How far are you from the source of the sound? The farther you are, the smaller the amplitude (see Lab 1.2).
3. What is the intervening material? For example, sound does not travel through walls as well as air.

Shape of the Repetitive Motion

Although a wave repeats in time, its motion during an oscillation can be simple or highly complex. For example, the following two graphs (Fig. 1.2) each show repetitive motion and the period and frequency of the motion is the same in both cases. We will be seeing quite a few graphs like this, so it is worth spending a moment to explain what is being shown. If you whistle into a microphone, you produce a fairly pure tone. The microphone responds by producing a voltage proportional to the sound pressure. This is what is shown in Fig. 1.2 (top graph). The voltage at the zero time is about 1.1 Volts (abbreviated V). It comes back to that value at around 1.6 ms, but the signal is not repeating yet, because here the voltage is increasing, not decreasing. At 2.3 ms, the signal is again 1.1 V and decreasing, so that is the time for one period of oscillation. If you look carefully at the bottom graph, you will see that it also has a period of 2.3 ms and, hence, the same frequency as the upper graph.

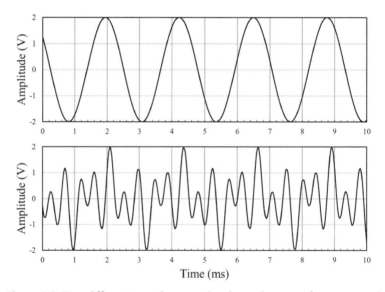

Figure 1.2 Two different sound waves that have the same frequency and period.

In fact, their amplitudes are also approximately the same. Thus, if these two waves represent sound waves, the pitch and loudness would be the same in both cases. But would they sound the same? The answer is no, because there is one more attribute to sound waves that you are familiar with, and that is its *tone quality*. This is what makes different instruments sound different. A violin and a trumpet can play the same pitch with the same loudness, but we can easily tell them apart, because they have a different tone quality. In fact, the same instrument can create different tone qualities. If you pluck a guitar in different ways, you can get quite different tones. The technical musical term for this is *timbre*. For general waves, we refer to the *shape* of the wave or sometimes the *waveform* (see Lab 1.3).

Worked Examples

1. Consider the Earth going around the Sun. Is this motion repetitive? What would an oscillation correspond to? What is the period and frequency of the motion?

Yes, this is a repetitive motion. Consider the moment when the Earth is closest to the Sun. We will consider that the starting point. Exactly one year later, the Earth will return to that point and this will happen year after year. So, the motion repeats. One oscillation corresponds to one trip around the Sun. The period is 1 year and the frequency is 1/year.

2. What is the period and frequency of the wave on the following graph?

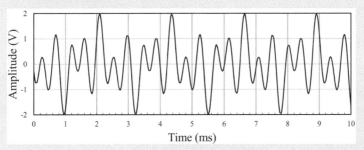

To find the period, we must identify a point on the wave that we can recognize and then find the time it takes for the wave to return to that same point. For example, we could consider the point where the wave reaches a maximum. The first time this happens is at 2.1 ms. The next time this happens is at 4.4 ms. So, the time for one oscillation, or the period, is 4.4 − 2.1 ms = 2.3 ms. The frequency will be $1/(2.3)$ ms = $1/(2.3 \times 10^{-3}$ s$)$ = 435/s = 435 Hz.

1.2   Units, Graphs, Formulas

One of the most difficult aspects of science is language. The language of science is initially a foreign language, as no one grows up from a young age speaking this language, making it difficult to learn. However, a precise language of science is critical to allow thousands and thousands of scientists to communicate with each across many different countries and cultures and across the ages. Ultimately, this textbook is a course in modern physics, but much of the early material concerns an ancient Greek scientist, Pythagoras. How is it that we can learn from Pythagoras? It is because he created a rational framework for describing and analyzing the natural world that is as valid today as it was 2000 years ago. This Pythagorean approach to the world has been so successful that we still use essentially the same approach today. Thus, understanding what Pythagoras did in ancient Greece will help us understand the endeavor of modern physics today.

As soon as we began to discuss various properties of sound waves in the last chapter, we started to introduce a new language. Part of this language is similar to our everyday language. Certain words are defined and given specific meanings, such as frequency. However, our scientific language included some new elements: units, graphs, and formulas. These elements are not used in everyday language and are less familiar. So, before we continue our discussion of waves, we must spend some time on these new elements.

Units

When we introduced the concept of frequency, we used the example of the note A 440. This raised the question, 440 what? The "what" is the unit. In this case, there are 440 oscillations/second. Thus, the unit is oscillations/second or Hz.

Any quantity must have a unit to go along with it. It makes no sense to say that it is 90 to Boston. One can say that it is 90 miles to Boston and, so, here, the unit is a mile. The only issue related to units is that there are often many different units for the same physical quantity. For example, it may be 90 miles to Boston, but it is also 475,200 feet or 149 kilometers or 158,400 yards. The physical distance is the same, but how it is expressed can change. So, we must be able to convert from one unit to another.

The key to converting units is to always multiply the original quantity by 1. This will not change the physical value. The only difficult part is to figure out what form of 1 to use. For example, we might want to know how many seconds there are in 1 day. We know that 1 day is the same as 24 h, so we can write:

$$1 = \frac{24\,\text{h}}{1\,\text{day}}.$$

Also, 1 h is equal to (60 × 60) s = 3600 s. So, $1 = \frac{3600\,\text{s}}{1\,\text{h}}$.

Finally, to convert 1 day into seconds, we have: $1\,\text{day}\frac{24\,\text{h}}{1\,\text{day}}\frac{3600\,\text{s}}{1\,\text{h}} = 86,400\,\text{s}$. The important point is that the days and hours cancel out, leaving just seconds. By cancelling out, we mean the following. Above, we introduced the idea that we can always multiply a number by 1, as it does not change the value, but we have to be creative about how we express the number 1. Here we have just the opposite: we can always remove a factor of one without changing the value of the number. In this case, we can write 1 = day/day, which means we can remove the day/day. Similarly, we can remove the hour/hour. Scientists call this cancelling out the units. Note that, in terms of units, hour and hours are the same thing.

Graphs

Graphs offer a way of presenting information to us that can make it easier to understand or extract information from a set of data. In fact, a graph can tell a whole story.

For example, a physicist works at a high-tech company and gets a raise of some amount each year. However, one year she makes a particularly important discovery and receives an extra raise. Her salary for each year is given in Table 1.1. In which year did she make the big discovery?

The first thing that you will probably notice is that the years are out of order. This makes it very hard to read the table. Second, even with a calculator, it would probably take you a while to figure out what year the big raise occurred. However, plotting the salary against the year makes it immediately obvious when the jump in salary occurred (Fig. 1.3).

Table 1.1 Yearly salary

Year	Salary ($)
2020	96,935.36
2019	96,772.67
2013	88,623.52
2017	93,881.00
2015	92,125.75
2012	88,210.00
2016	93,387.34
2018	95,257.21
2010	85,814.16
2014	89,393.14
2011	87,120.89

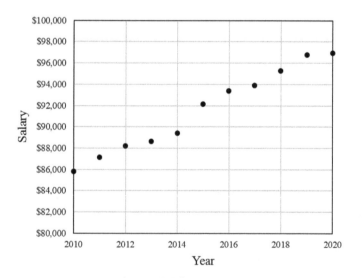

Figure 1.3 Salary vs. year.

For another example, the following graph shows my progress as I drove to Boston one day (Fig. 1.4).

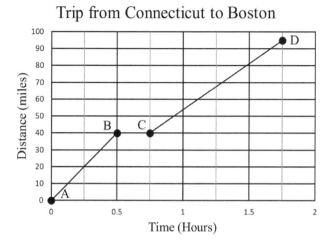

Figure 1.4 Trip from Connecticut to Boston.

This graph tells a story—the goal is to explain what hasppened. At first, your explanations may seem contrived. But, when you figure it out, you really feel that you have the right answer. Actually, there are at least two or three good answers. Can you find them all? In fact, you can even tell something about when the problem was written.

Formulas or Relationships

Formulas or relationships are probably the hardest part of the language of science for novice scientists to use or appreciate. However, they are probably the most important result from scientific work. They succinctly contain and summarize the information and knowledge that physicists produce. Moreover, the meaning and significance of formulas is never really taught to novice-scientists, so we will spend quite a bit of time on them.

For example, the frequency at which a string can vibrate is given by the following formula:

$$f = \frac{1}{2L}\sqrt{\frac{T}{U}}. \tag{1.2}$$

Here, L is the length of the string, T is the tension on the string, and U is the density of the string. We will return to this formula many times in the next chapters.

Or, another formula that you might remember from a High School science course that describes a gas in a container is:

$$pV = nRT, \tag{1.3}$$

where n is the number of moles of gas, R is the gas constant, T is temperature and V is the volume of the container. If you are given the values of the various quantities in the formula, you can readily type the numbers into a calculator and get a result. However, formulas contain much more information than this. For example, perhaps we don't know the value of all the quantities in a formula. In this case, you might think that the formula will be useless. On the contrary, it can still provide important information, as we will see below.

In many physics courses, a lot of time is spent deriving formulas, which can often involve more advanced mathematics. The derivation of many formulas requires calculus, so we will not concern ourselves with derivations. This is not unreasonable: theoretical physicists devote their time to deriving formulas, and, often, I, as an experimentalist, cannot even understand the mathematical methods that they use. But this is not a problem, as long as I can both *understand how to use* the formula and *test* it to make sure that it is correct. A physicist can use very fancy mathematics, but if the formula does not work in an experiment, it must be wrong. Every concept and dependence in this course can be tested with the labs outlined in the Appendix. For another example of dependences, see Lab 1.4.

1.3   Dependencies: Qualitative and Quantitative

There are four steps to fully understanding a formula. We will illustrate these steps with the formula for a gas in a container mentioned above.

The question is the following: What is the pressure of the gas in a container?

Step 1: What does the quantity of interest *depend* on? Here, one must be creative and just try to think of everything that might affect the quantity of interest. For our example, these would include the volume of the container, the temperature, the number of gas particles, the type of container, type of gas, the altitude, etc.

Step 2: Find or derive a formula that will connect these quantities together. Solving Eq. 1.3 for *p* yields:

$$p = \frac{nRT}{V}. \qquad (1.4)$$

This is the hard part of physics. Even researchers cannot always understand how another scientist came up with a formula. No scientist can understand, in detail, everything every other scientist has done. That would be simply impossible. But, what a scientist must be able to do is use the formula correctly and test it in the lab. That is the approach we will take in this course.

Step 3: Qualitative check.

The first check of a formula is relatively straightforward: if you change one of the variables in the equation, does the quantity of interest change in the correct way? For example, if the temperature in a container is increased, how should the pressure change? Think about (but don't try!) throwing a used spray can on a fire. It is pretty clear that the pressure in the can will increase until the can explodes. Does this agree with the formula above? If you increase *T* in the expression (Eq. 1.4) and keep *everything else* (*n*, *R*, and *V*) *the same*, *p* will also increase. Thus, the formula *qualitatively* agrees with what you know to be true. If it did not, you would know immediately that the formula was incorrect. Now, you can go on and check the rest of the variables: for example, if I increase the volume, should the pressure increase or decrease and does the formula give the correct result?

Step 4: Quantitative check.

The quantitative check involves some mathematics and can be a little trickier, but like any skill, all that is needed is some practice. Here, we are asking a more detailed question than in Step 3, namely, if I change one variable by some factor, by what factor will the result change? Again, it is very important to keep all the other variables the same. This is a quantitative question, because now we need a numerical answer. For example, if I double the temperature (i.e., increase the temperature by a factor of 2) how will the pressure change? This is not a question that you can answer, right off. So, first, we must learn how to extract an answer from the formula. Second, we must do an experiment to find out whether this answer is correct. How one quantity (pressure) changes with another (temperature) is called a *dependence*; in other words, it answers the question how one thing depends on another.

Calculating Quantitative Dependencies

Consider the pressure of a gas in a container given by Eq. 1.4. Notice that you might not understand some of the terms in the equation, like the number of moles of gas or the gas constant. However, the great thing about dependencies is that you do not need to know or understand everything to answer some useful questions. This is a really important goal of physics, to make interesting observations or predictions even in the face of incomplete knowledge.

A typical problem starts with knowing the pressure in the container for particular values of n, T, and V. If we change one of these quantities, we want to know how the pressure will change. For example, the pressure is 35 pounds/square inch (PSI) (about the air pressure in a typical car tire). What is the new pressure if I increase the temperature by a factor of 1.25?

Here is how to proceed: the original, or known, pressure is p_{known} = 35 PSI. So, from Eq. 1.4, $p_{known} = \dfrac{nRT}{V}$. The new pressure will be: $p_{new} = \dfrac{nR(1.25T)}{V}$ because we have replaced T with $1.25T$. In other words, we have increased the temperature by a factor of 1.25.

The next step is to isolate the factor of 1.25 that you introduced into the formula: $p_{new} = 1.25\left(\dfrac{nRT}{V}\right)$. What is left in the bracket is just the known frequency, p_{known}, so we can substitute that into the expression for the new pressure: $p_{new} = 1.25p_{known}$. Finally, the new pressure is $p_{new} = 1.25(35\text{ PSI}) = 43.75\text{ PSI}$.

We will now consider an example that is much more important for this class: the frequency of a vibrating string.

Step 1: What does the frequency of a string depend on?

The length of string, L, the tension or how tightly it is pulled, T, and the density or thickness of the string, U.

Step 2: What is the formula for the frequency of the string? At this point, we don't know, but let's consider the following:

$$(a)f = \frac{L}{UT}; (b)f = \frac{4}{L^2}TU; (c)f = \frac{1}{2L}\sqrt{\frac{T}{U}};$$

$$(d)f = \frac{1}{L}\sqrt{\frac{U}{T}}; (e)f = \frac{T^2}{U\sqrt{L}}; (f)f = 2T\sqrt{UL}$$

Step 3: Qualitative dependence.

We know that if you increase the length of the string, the frequency will decrease, so this rules out (a) and (f). If you increase the tension, the frequency will increase. This rules out (a) and (d). If you increase the density of the string, the frequency will decrease. This rules out (b), (d), and (f). This leaves only (c) and (e). You will need a quantitative test to choose between these two, but you do not have to worry about the others.

Step 4: Quantitative dependence.

It turns out that Eq. 1.2 is correct. So, now we can see what happens to the frequency if we change the length of the string.

If we know the frequency of a string is 500 Hz and we triple the length of the string, what is the new frequency? Following the same procedure as we did above for pressure, we start with $f_{known} = \frac{1}{2L}\sqrt{\frac{T}{U}}$. To find the new frequency, we replace L with $3L$ giving $f_{new} = \frac{1}{2\cdot 3L}\sqrt{\frac{T}{U}}$. We again need to isolate the new factor: $f_{new} = \frac{1}{3}\left(\frac{1}{2L}\sqrt{\frac{T}{U}}\right)$. Finally, we rewrite this in terms of the known frequency: $f_{new} = \frac{1}{3}(f_{known})$. Putting in numbers, we get $f_{new} = \frac{1}{3}(500 \text{ hz}) = 167 \text{ Hz}$.

Questions on quantitative dependences can also be asked in the reverse way and this is sometimes more important. For example, we can ask, by what factor do I have to change the length of a string to make its frequency increase by a factor of 2? One way to solve this is to rewrite the relevant formula to show how length depends on frequency, in this case: $L = \frac{1}{2f}\sqrt{\frac{T}{U}}$. Then the procedure is the same as before. Or you can make a guess as to what change in the length will work and try it out. If it does not give the correct result, you can usually then predict what will work. The nice thing is that you can always check your answer to make sure it works.

Experiments and Formulas

The last aspect of formulas, briefly mentioned above, is verifying them experimentally. This is important when you have a formula, but you don't know from where it came. Rather than just accepting it as correct, you can test it in the lab.

For example, we want to test the formula for the frequency of the string. It is hard to test the whole formula at once, so we will look at one part of it: the dependence of frequency on the length of the string. Interestingly, we can do this without knowing anything about the other quantities in the formula. Once again, consider Eq. 1.2. Since we are interested in frequency and length, we can move them to the same side of the equation:

$$fL = \frac{1}{2}\sqrt{\frac{T}{U}}. \qquad (1.5)$$

The important thing to notice is that the product $f \cdot L$ is constant if we hold the tension and density of the string constant. The value of the constant does not concern us; we just note that it is a constant and this gives us a way to test the dependence. We set up a vibrating string in the lab and measure the frequency for different values of the length of the string (while keeping everything else the same) shown in Table 1.2 (see also Lab 2.1):

Table 1.2 Sample data for the dependence of the frequency of a string on length

L	f	fL	f/L	fL^2	$f\sqrt{L}$
1.05 m	478 Hz	501.9 Hz-m	455.2 Hz/m	527.0 Hz-m^2	489.8 Hz-m$^{1/2}$
0.51 m	1005 Hz	512.6 Hz-m	1971 Hz/m	261.4 Hz-m^2	717.7 Hz-m$^{1/2}$
1.98 m	252 Hz	499.0 Hz-m	127.3 Hz/m	987.9 Hz-m^2	354.6 Hz-m$^{1/2}$
	Max/Min:	1.02	15.5	2.78	2.02

From this example data, we can calculate various combinations of f and L. If we find a combination that is, at least, approximately constant, even as we change the length of the string, we know that we have found out how the frequency depends on the length of the string. To determine which column is the most constant, we can divide the maximum value in the column by the minimum value. Then choose the column whose ratio is closest to one. In this case, it turns out that $f \cdot L$ = constant, and so, f = constant/L, as it should. However, we did not have to rely on the formula to reach this conclusion. We could have learned this simply from the data.

Worked Example

As we will see in Chapter 3, the frequency of a vibrating bar, like on a xylophone, has a different dependence on length than a string. This greatly affects the construction of a xylophone. It turns out that the frequency of a bar is $f = \dfrac{c}{L^2}$, where c is a constant for a particular kind of bar and L is its length. Even though we don't know what c is, we can answer quantitative questions. Assume we have a bar that is 20 cm long and vibrates at a frequency of 235 Hz. What will be the frequency of a bar 27 cm long?

To start, we note that we have made the bar (27 cm/20 cm) = 1.35 times longer. We can also write that $f_{known} = \dfrac{c}{L^2}$ = 235 Hz. Then,

$$f_{new} = \frac{c}{(1.35L)^2} = \frac{c}{(1.35)^2 L^2} = \frac{1}{(1.35)^2} f_{known}$$
$$= 0.549 \times 235\,\text{Hz} = 129\,\text{Hz}.$$

The experiment in Table 1.2 does not tell us everything about the frequency of a string, but we do learn something. We can also vary the tension in the string and keep the length and thickness constant to study how the frequency depends on the tension of the string.

Conclusion

The most significant conclusion from this section is that we do not need to know or understand every aspect of a formula to be able to obtain some useful information from it. We can answer certain questions without knowing everything. If it was not for this, physics and science would never have progressed as there are always unknowns.

1.4   Problems

Section 1.1

1.1   Briefly describe the difference between pitch, timbre, and loudness in terms of sound waves.

1.2   (a) What is the period of this wave?
       (b) What is the frequency of this wave?

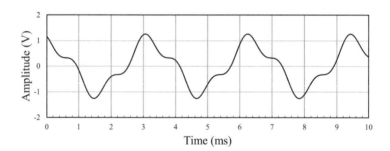

1.3   (a) What is the period of this wave?
       (b) What is the frequency of this wave?

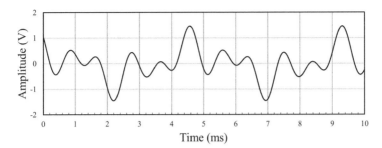

1.4 Consider the following graphs:

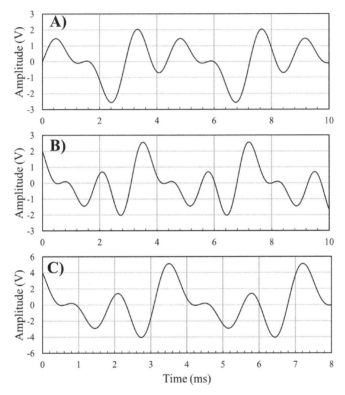

(a) Which wave has a different frequency from the other two?

(b) Which wave is the loudest—i.e., largest amplitude?

(c) Which wave has a different shape from the others?

(d) Which is not a wave?

Answers: A: A); B: B); C: C); D: All of them; E: None of them

1.5 Which of these factors affects the loudness of a sound?

(A) Distance from the source of the sound.
(B) Pitch of the sound.
(C) Temperature of the air.
(D) Shape of the wave.

1.6 Before CDs, people listened to records on a turntable. If the turntable revolved at 45 revolutions per minute, what is the period of this system in seconds? What is the frequency in Hz?

Section 1.2

1.7 Match the following graphs with their functions:

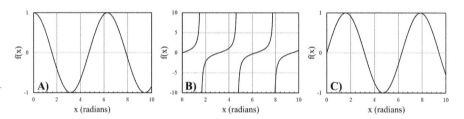

a: $\tan(x)$; b: $\cos(x)$; c: $\sin(x)$

1.8 Read off the value of $f(x)$ for $x = 7.3$.

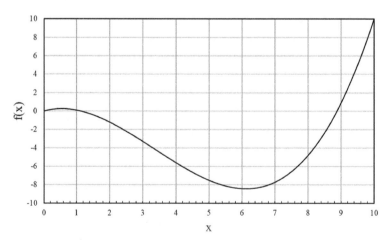

1.9   (a) Convert 55 miles/h to ft/s:

(b) Convert 343 m/s to miles/h:

Section 1.3

1.10   What does the maximum speed of your car depend on?

1.11   The frequency of the fundamental of a string is given by Eq. 1.2.

(a) If I increase the length, will the frequency increase or decrease?

(b) If I increase the tension, will the frequency increase or decrease?

(c) If I increase the string density, will the frequency increase or decrease?

(d) If I change the length by a factor of 5, by what factor does the frequency change?

(e) If I change the tension by a factor of 5, by what factor does the frequency change?

(f) If I change the string density by a factor of 5, by what factor does the frequency change?

1.12   The frequency of the fundamental of a string is given by Eq. 1.2. I start with a frequency of 1700 Hz. If I decrease the tension by a factor of 25, what will be the new frequency?

1.13   (a) If you increase the density, U, of a string, the frequency will (increase | decrease | stay the same)?

(b) What has a greater effect on the frequency of the string: Doubling the density or doubling the length?

(c) Originally, the string has a frequency of 375 Hz. If I increase the tension by a factor of 2.4, what is the new frequency of the string?

1.14   The frequency of the fundamental of a string is given by Eq. 1.2

(a) If I double the length of the string, how will the frequency change?

(b) If I increase the density of the string by a factor of 16, how will the frequency change?

(c) If I increase the tension by a factor of 9, how will the frequency change?

(d) If I increase the tension by a factor of 3, how should I change the density to keep the frequency the same?

(e) If I increase the density by a factor of 4, how should I change the length to keep the frequency the same?

1.15 The frequency of the fundamental of a string is given by Eq. 1.2.

I start with a frequency of 700 Hz. If I increase the tension by a factor of 4 and increase the density of the string by a factor of 9, what will be the new frequency?

1.16 The period of the orbit of a planet around the Sun is proportional to $\sqrt{\dfrac{R^3}{M_{Sun}}}$. R is the distance from the Sun and M_{Sun} is the mass of the Sun.

(a) If I decrease the distance, will the period increase or decrease?

(b) If I increase the mass, will the period increase or decrease?

(c) If I change the distance by a factor of 5, by what factor does the period change?

(d) If I change the mass by a factor of 5, by what factor does the period change?

1.17 The period of the orbit of a planet around the Sun is given in problem 1.16.

(a) If the Earth was twice its distance from the Sun how long would it take the Earth to circle the Sun?

(b) If the Sun was three times as massive, how long would it take the Earth to circle the Sun?

1.18 The speed of a wave on a string is $v = \sqrt{\dfrac{Mg}{U}}$, where M is the mass on the end of the string, g is the acceleration of gravity and U is the mass per unit length of the string. For a particular string, a mass of 1 kg gives a velocity of 100 m/sec. What mass would produce a velocity of 300 m/sec?

1.19 The density of helium is 7 times smaller than air. For any gas, the speed of sound is given by: $v = \sqrt{\dfrac{\text{Pressure}}{\text{Density}}}$. If the pressure is kept the same, what is the speed of sound in helium, if the speed of sound in air is 343 m/s?

1.20 The density of xenon is 4.7 times larger than air. For any gas, the speed of sound is given by: $v = \sqrt{\dfrac{\text{Pressure}}{\text{Density}}}$. If the pressure is kept the same, what is the speed of sound in xenon, if the speed of sound in air is 343 m/s?

1.21 The speed of a wave on a string is given by: $v = \sqrt{\dfrac{\text{Tension}}{\text{Density}}}$. A particular string gives a speed of 400 m/sec. If we want to increase the speed to 600 m/sec, by what factor should we change the density of the string, if we keep the tension the same?

1.22 The speed of sound depends on the air pressure and the air density in the following way: $v = \sqrt{\dfrac{\text{Pressure}}{\text{Density}}}$. We can change the density of air by changing its composition.

(a) In normal air the speed of sound is 343 m/sec. If I increase the density of the air by a factor of 3, does the speed of sound increase or decrease? By what factor will it increase or decrease?

(b) If I want to increase the speed by a factor of 4.5, how should I change the density of the air?

1.23 Determine how the frequency, f, of a pendulum clock depends on the length, L, of the pendulum using the following data:

S.N.	Length, L	Frequency, f	fL	$f\sqrt{L}$	f/L
1	0.1 m	1.576 Hz			
2	0.3 m	0.910 Hz			
3	0.8 m	0.557 Hz			
4	1.2 meter	0.455 Hz			
5	1.9 m	0.361 Hz			

1.24 Graph the functions $f(x) = \cos(x)$ and $g(x) = \sin(x)$, measure x in radians. Graph the functions $h(x) = x^2/10$ and $k(x) = \sqrt{x}$.

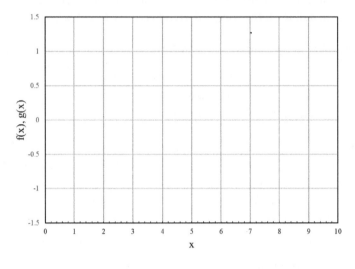

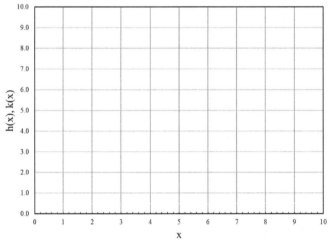

Chapter 2

Melodic Instruments and the Origins of Music and Science

2.1 Neanderthal Flutes and Greek Lyres

The origins of music and science go far back in time, perhaps even earlier than modern human beings. More importantly, the connection between music and science appears to be equally ancient. We will explore this history with two instruments. The Greek lyre is a stringed instrument that was thoroughly studied by the Greek mathematician, Pythagoras in about 600 BC. The second instrument was only recently discovered and dates back to the time of the Neanderthals and appears to be a flute made out of an animal bone.

Both the flute and the lyre are fairly sophisticated instruments upon which a melody can be played and thus are called melodic instruments. The other main group of instruments is called percussion, which includes drums, cymbals, and triangles. They can be used for great dramatic effect in a piece of music but cannot sustain a melody. Xylophones fall in between melodic and non-melodic, but we will consider them to be percussion instruments for reasons that will become clear in Chapter 3. One might think that percussion instruments are easier to make than melodic instruments and would have been developed earlier in human history. While this may be true, the physics of melodic instruments

The Music of Physics: An Introduction to the Harmonies of Nature
George N. Gibson
Copyright © 2024 Jenny Stanford Publishing Pte. Ltd.
ISBN 978-981-4968-99-7 (Hardcover), 978-1-003-48481-3 (eBook)
www.jennystanford.com

is actually simpler than that of percussion instruments. In light of this, it is not surprising that the *science* of musical instruments started with melodic instruments.

For this same reason, we will also begin our study of music and instruments with melodic instruments and only later delve into the more complex physics of the percussion instruments.

The Neanderthal Flute

Not too long ago, a remarkable discovery was made: an animal bone was found at a Neanderthal camp sight. The significant feature of the bone is that it has fairly small very circular holes drilled into it. It obviously took some time and skill to create these holes, which immediately raises the question: why did the Neanderthals spend the effort to make them? The pattern of the holes is also striking: they are in a line but they are not equally spaced. This bone can be seen at: https://en.wikipedia.org/wiki/Divje_Babe_flute.

Although the conclusions are controversial, it appears that the bone is, in fact, a musical instrument as the arrangement of the holes may be the same as that of a modern flute. If correct, this is a profound insight because of the many implications:

- Music has been around for a very long time and predates modern humans.
- The Neanderthals created and appreciated music.
- They were capable of skilled craftsmanship.
- They used their technical skills to make musical instruments.
- They had some level of reasoning to choose a particular hole spacing.
- They may have used the same harmony and musical scale as we use today.

This last point is as subtle as it is significant and we will need to learn a bit more about musical scales and instruments before we can fully appreciate it. However, it does hint at the conclusion that Western music and harmony is not a purely cultural creation but has its roots in some basic physics that applied to the Neanderthals just as much as it applies to us. It also encourages us to examine more closely the connection between physics and music.

Pythagoras and Greek Lyres

The idea that it might be interesting to apply scientific analysis to music and musical instruments is not particularly new. In fact, it was thought of at least back to the time of Pythagoras. Pythagoras was, as far as we know, the first person to seriously make connections between mathematics, science, and music. Most everyone has been exposed to the Pythagorean Theorem sometime in high school. This is just one of the many results and accomplishments of Pythagoras and his musical contributions are less well known.

The popular instrument of the day in Pythagoras' time was the lyre. It consisted of a frame on which were strung four strings and each string was adjusted to produce a certain pitch when plucked. Pythagoras was intrigued by this instrument and wanted to understand exactly why the strings were set at particular pitches and why these pitches followed a particular pattern. However, before we continue, we must introduce a bit of music theory—about the only music theory that we will need for the rest of the course.

Musical Scales

When we start to examine music and music theory, one of the most significant observations is that not all pitches are used. Consider a violin: you can move your figure along the string to produce a continuous range of pitches. However, in any musical composition, you only play from a certain group of pitches called notes. Depending on the type of music, Western, Arabic, Asian, Indian, etc., there may be many or few notes that are used, but the point is that there is a definite collection of pitches that are played. For special effects, musicians may slide from one note to another covering a continuous range of pitch, but this is not really part of the melody or harmony. This set of notes is organized into what is called a musical *scale*. Again, the particular notes that comprise a scale vary from culture to culture, but organizing the notes into a scale seems to be universal.

All scales share another feature: they are specified only over a small range of pitches called an *octave*. Most everyone is familiar with the names of the notes on the piano: C, D, E, F, G, A, B, C (see Fig. 2.1). These notes are the white keys on the piano, and anyone who has taken piano lessons probably started out by playing this sequence. However, there are more than eight white keys on the piano, so what are the names of the rest of the notes? We will come back to this question later. If you ever look at a piano keyboard, you would notice that the pattern of white keys and black keys repeats up and down the piano. In the same way, the names of the notes repeat up and down the keyboard. With this in mind, the sequence of notes above form a scale, called the C major scale. Once you reach the end of the scale, it simply repeats as you go higher or lower. Furthermore, the range from one C to the next is called an octave. We will discuss all of this in much more detail in the next few sections, but it is helpful to get some sense of what is coming up.

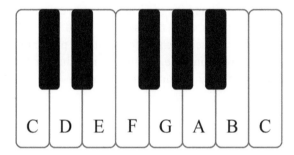

Figure 2.1 Keys covering one full octave on a keyboard instrument.

Returning to Pythagoras, as mentioned above, he was intrigued by the fact that the musical scale exists and he wanted to know why Greek music was based on a particular kind of scale, in other words, why did Greek musicians use a particular sequence of notes for their musical scale? These questions that Pythagoras asked launched a way of thinking and analyzing the world that has produced a tremendous body of knowledge that Pythagoras could never have dreamed of. Indeed, Pythagoras' method of understanding the musical scale eerily foreshadowed the sequence of steps that led to the development of quantum mechanics and modern physics, many centuries later.

2.2 Pythagorean Intervals

As discussed in the previous section, Pythagoras was interested in understanding the notes and scales used in Greek music. In particular, he studied the Greek lyre, described above. His first observation was that if you have two strings with the same length, tension and thickness, they sound the same when you pluck them. This means they have the same pitch and sound good (or *consonant*) when played together. The second observation was that if the strings have different lengths (keeping the tension and thickness the same) the strings have different pitches and generally sound bad (or *dissonant*) when played together. Finally, he noticed that for certain lengths, the two strings still had different pitches, but now sounded consonant rather than dissonant. Pythagoras called the relationship between two notes an *interval*.

For example, as mentioned above, when two strings have the same length, they have the same pitch, and the relationship, or interval, between the notes is called a *unison*. If one string is exactly one-half of the length of the other string, it pitch is much higher, but they still sound consonant when played together. This interval is called an *octave*. Finally, if one string has a length that is two-thirds the length of the other, the strings again sound consonant when played together and this interval is called a *perfect fifth* (the reason why this is called a fifth and a perfect one at that will be discussed in a little bit).

Already, we have a very important distinction. These internals are defined by the lengths of the strings being in a certain *ratio*. One does not form an octave by reducing the length of a string by a fixed amount, like 10 cm. Rather, one forms an octave by *dividing* the length of a string by a factor of 2. In other words, if the lengths of the strings are in a ratio of 2 to 1, the pitches of the strings will form an interval of an octave.

Musically, there are many different intervals, each with its own characteristic sound and each with a special role in music harmony. We will encounter quite a few intervals during the course. Each interval will have a name, like an octave, and it will be defined by a ratio, like 2:1 for the octave.

So far, our list of intervals includes:

Name	Ratio
Unison	1:1
Octave	2:1
Perfect fifth	3:2

As mentioned above, each interval has a particular characteristic. The unison may not seem so interesting, as the two notes are the same. This may be true, but it is the simplest and most fundamental interval and it can be used to great effect. Imagine a singer singing a simple melody. Now imagine a chorus of a hundred voices, all singing the same melody in *unison*, i.e., everyone singing exactly the same pitch. It is actually very hard to have that many people sing perfectly enough, but when they do the effect is quite dramatic. So, great composers can use even the simplest interval effectively.

The octave is the next most important interval. As discussed above, it defines the range of the music scale. Two notes an octave apart sound so similar that they are given the same name. For example, elementary piano pieces often start on Middle C. However, if you go up an octave from there, the note is still called a C. In fact, it is sometimes hard to tell whether two notes are one or two octaves apart. If someone has a deep voice and cannot sing a particular melody because the notes are too high, it is generally acceptable to sing the melody an octave lower. Musically, the melody is the same.

Finally, we come to the perfect fifth. In this case, there is no question that two notes a perfect fifth apart are really two different notes. On the piano, the note a perfect fifth above Middle C is called a G. Thus, the perfect fifth is the first interval to get us into *harmony*, the question of how different pitches interact with each other. Once again, the perfect fifth has a special quality to it—anyone who has heard Gregorian chant will recognize the open haunting quality of the perfect fifth. Looking back at Fig. 2.1, we see that the perfect fifth is the fifth note of the scale.

So far, we have only talked about intervals in terms of going *up* in pitch. However, we can go down, as well. If you start with a string and you double its length, the pitch will go *down* by an octave.

String 2	1/2	2	2/3	3/2
String 1	1	1	1	1
Ratio	1/2	2/1	2/3	3/2
Interval	Up an octave	Down an octave	Up a Fifth	Down a Fifth

It is also very important to notice that intervals do not depend on the initial frequency, length, tension, or type of string. Intervals only give a relationship between two notes, but so far, like Pythagoras, we only defined them in terms of lengths of strings. It turns out that for a string, if you divide the length of the string by a factor of two, the frequency goes up by a factor of two (review Section 1.3 to understand why). So, whether we are considering the lengths of the strings or their pitches, the ratios for the intervals are the same.

Musically, intervals are very important, but why are we discussing them so carefully in a physics course? From a scientific standpoint, by far the most important characteristic of these simple consonant intervals is that they are defined by ratios of small integers. This was very significant to Pythagoras. Pythagoras thought about mathematics and arithmetic, and the properties of integers, in particular. But, this was just abstract thought. Pythagoras had no way of knowing whether mathematics would actually be useful for anything. It then turns out that integers and simple ratios could describe the consonant musical intervals.

This also highlights a deep connection between the human experience and abstract mathematics. What appears to be a subjective judgment, this interval sounds good (consonant), that interval sounds bad (dissonant) can actually be predicted using abstract mathematics. Apparently, our emotional response to the world sometimes follows mathematical laws.

Returning now to intervals, we can define a new operation, namely, combining intervals. For example, we could go up in pitch by an interval of an octave and down by an interval of a perfect fifth. In this case, going up by an octave means multiplying the frequency by a factor of 2. Going down by a fifth means dividing by 3/2. All together, we have 2/(3/2) = 4/3. Thus, by combining intervals, we have produced a new interval, called the perfect fourth—the fourth note of the scale. The perfect fourth is defined by a ratio of 4/3.

To summarize:

- Ratios of 1/2 and 2/1 give octaves.
- Ratios of 2/3, 3/2 give fifths.
- Ratios of 3/4, 4/3 give fourths.

Notice that the ratios above only involve the integers 1, 2, 3, and 4. As an exercise, it is helpful to write out ALL ratios involving these integers:

1/1	1/2	1/3	1/4
2/1	2/2	2/3	2/4
3/1	3/2	3/3	3/4
4/1	4/2	4/3	4/4

Now, try to determine what interval each of these corresponds to. For example, 2/3 corresponds to going down by a perfect fifth. 4/1 corresponds to going up by an octave *twice*. In other words, since going up an octave means multiplying by 2, going up two octaves means multiplying by 2 twice, or 2 × 2 = 4.

At this point, we can ask, why stop at 4? Why not use 5, 6, 7, and 8 as ratios for intervals? The answer is that we can use larger integers. However, Pythagoras stopped 4 because he realized that he had all the intervals that he needed to construct the musical scale, and that was his original goal. Moreover, Pythagoras was a great believer in simplicity—he wanted the simplest explanation for the musical scale possible. He believed that the simplest explanation must be the correct one. To this day, this continues to be a central conviction in physics and science.

In fact, Pythagoras was so pleased with these intervals that he called them "perfect", i.e., the perfect fifth and the perfect fourth. (The unison and octave are also perfect, but as discussed above, they are so perfect they are not really considered to be a new note.)

In a later section, we will discuss how to create a musical scale from these intervals, but first we will organize what we have learned in a slightly different way, which will make constructing the scale a bit easier.

Table 2.1 shows the relationship between notes on the scale, intervals and frequencies, starting with the note D, using 293 Hz as a standard choice for D:

Table 2.1 Four intervals starting with the note D

Name of note	Name of interval	Ratio of interval	Decimal equivalent	Frequency of note
D	Octave	2/1	2.000	587 Hz
A	Perfect fifth	3/2	1.500	440 Hz
G	Perfect fourth	4/3	1.333	391 Hz
D	Unison	1/1	1.000	293 Hz

Although we don't yet have enough notes for a complete musical scale, these are, in fact, the notes to which the strings on the Greek lyre were tuned. Also, Gregorian chant was mostly based on fourths and fifths and, so, used these notes primarily. Of course, we can start on any note we want, D = 293 Hz is just an example.

2.3   Vibrating Strings

Pythagoras started his studies of musical scales by noting that two strings whose lengths formed certain ratios sounded pleasing when played together. How was this done in practice? One can make a two-stringed instrument and adjust the tension on the strings so that they have the same pitch (an interval of a unison) (Fig. 2.2).

Figure 2.2 Simple two-stringed instrument to demonstrate intervals.

On most string instruments like this, the pitch is changed as one plays, by placing a finger on the string and pressing down hard. This restricts the length of string that can vibrate, which then raises the pitch. If you press one string, say string 2, half way along its length (at point A), it will vibrate like a string half of its length and its frequency will be a factor of 2 higher. This new pitch will then form an octave with respect to string 1.

To form a perfect fifth between the two strings, we want the frequency of string 2 to be a factor of 3/2 higher, or its length should be a factor of 2/3 shorter. So, we want to place our finger 2/3 the distance from one end, or 1/3 the distance from the other end. This puts us at point B. As long as the longer piece of the string is vibrating, the pitch will now be a perfect fifth higher than string 1.

To form a perfect fourth, the frequency of string 2 should be 4/3 higher, or the length 3/4 shorter. Thus, we can place our finger 3/4 the distance from one end, or 1/4 the distance from the other end. Again, we assume that the longer piece of the string vibrates.

All in all, we find that there are special points on a vibrating string such that when we press the string at those points the string vibrates at a frequency that forms a pleasing interval with respect to the identical string, as seen in Fig. 2.3.

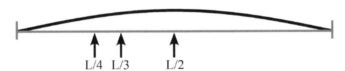

L/4 L/3   L/2

Figure 2.3 Points on the vibrating string that create an octave, fifth, and fourth with respect to the original string.

If the string has a length of L, these special points occur at a distance L/n from one end of the string, where n is an integer. We could also say that they occur at a distance $(n - 1)L/n$ from the other end of the string. As it turns out, these points at L/n are special for another reason, as well.

Rather than press down hard with our finger, we can just very lightly touch the string at some point and pluck the string with our other hand. Usually, doing this does not produce a good sound, even though the whole string can vibrate. However, at certain points, a nice tone is produced with a much higher frequency than the original string. Moreover, the points where a nice tone is made are exactly the same points that we found above, namely a distance of L/n from one end of the string. In the case where we press down on the string, we essentially formed a somewhat shorter string, but nothing else changed. However, when we just touched the string, the whole string could vibrate, yet it produced a new pitch. In fact, after we pluck the string, we can remove our finger, and the string continues to vibrate at the new frequency. This is something new—it means that a string can vibrate at more than one frequency, but only certain new frequencies. The basic, or lowest frequency at which a string can vibrate is called the *fundamental*. In fact, the lowest frequency that anything can vibrate at is called its fundamental and the frequency is generally called f. All other frequencies that a string can produce are called *overtones*. Pythagoras based all of his work on the fundamental frequency of a string and he may not have known that a string also has overtones.

Although we now know that a string can vibrate at more than one frequency, it cannot vibrate at *any* arbitrary frequency. So, the question now is: what frequencies can a string vibrate at? Before we try to answer this question, it is useful to consider exactly what a string looks like while it is vibrating. We know that the endpoints of a string do not move, as they are the places where the string is attached to the instrument. A point on a string that does not move is called a *node*. All strings, on musical instruments, at least, have nodes at each end, but *they can have additional nodes along the string*. Here is what a simple string looks like with just nodes at the ends (Fig. 2.4).

● Nodes

Figure 2.4 String vibrating at its fundamental frequency.

What happens if we try to add another node right in the middle of the string?

In this case, notice that each part of the string is exactly half as long. The tension in the string and its composition has not changed, but a string half as long has a vibrational frequency that is twice as high. So, if the simple string has a frequency of f, the string with an extra node in the middle has a vibrational frequency of $2f$.

What happens if the node is not exactly at the center of the string?

Now the string is divided into two unequal sections. Each section would behave as a separate string, but since the lengths of the strings are different, their vibrational frequencies would be different. In this case, the whole shape of the string would not repeat with a single frequency. So, if only one node is added, it must be at the center of the string. In general, any number of nodes can be added to a string as long as each section of the string has the same frequency. In other words, each section of the string must have the same length.

If nodes are the points on the string that do not move as the string vibrates, then the points on the string that move the most are called antinodes. The antinodes are exactly half way between the nodes, as seen in Fig. 2.5. The frequencies and shapes of the modes for a vibrating string can be measured in Lab 2.1.

Each of these different patterns of nodes along the string is referred to as a *mode of vibration*, each with its own vibrational frequency. To determine the frequency, we just have to figure out the frequency of the shortest section between two nodes. For example, if the string has two extra nodes, the shortest section of the string has a length of $L/3$. Since the frequency of vibration is inversely proportional to the length of the string, a string with a length of $L/3$ will have a frequency of 3 times the fundamental frequency, or $3f$. In general, the short section of the string has a length of L/n, if there are $(n-1)$ extra nodes. So, the overtone would be nf. The whole series is then $1f, 2f, 3f, 4f, 5f, ...$

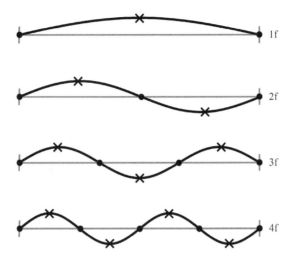

Figure 2.5 First four modes of a vibrating string with the nodes (•) and antinodes (x) marked.

The set of frequencies at which a string or any system can vibrate is called an *overtone series*. There are many systems, such as a string, which have a simple overtone series, in which all of the overtones are integer multiples of the fundamental. This is called a *harmonic series*. Finally, it sometimes happens that a system vibrates only at harmonics of the fundamental, but not all harmonics are possible, for example, the series *f*, 3*f*, 5*f*, ... This is called a *partial harmonic series*. That is, this is a harmonic series with some of the harmonics missing.

An interesting question arises from this analysis: how exact are the integers? In other words, if the fundamental has a frequency of 350 Hz, is the second harmonic *exactly* 700 Hz? In fact, the harmonics are nearly perfect integers. The concepts of precision and accuracy are rather hard to define, but there is a simple way to see how accurate the harmonics are. Imagine trying to pick a point exactly one third of the way from one end of the string without using a ruler—just by eye. You probably won't be very accurate. Now imagine trying to find a point one third of the distance from the end by listening for a clear harmonic. When you find this point, you will have divided the string into three parts far more accurately than you did by eye. This shows that there are different ways to measure things and often, if you base the measurement on frequencies, the measurement will be the most accurate. This is the concept, or technique, of *spectroscopy*, which is a major theme of this book.

One final note for this section: in modern physics, this idea that a system has only special, very specific, modes of vibration, each with a particular vibrational frequency is called "quantization". Each mode of vibration is called a quantum state and the properties of these states are understood through quantum mechanics. It was the realization that objects like electrons and protons are actually waves and have modes of vibration that led to the rapid development of modern physics in the early 1900s. Quantum mechanics made possible almost all modern technologies in one way or another. For example, computers, nuclear reactors, lasers, and magnetic resonance imaging (MRI) are all based on quantum mechanics.

2.4   Pythagorean Scales

Intervals to Scales

In Section 2.2, we introduced the concept of intervals and how one goes up and down by a certain interval. However, Pythagoras' real goal was to explain the musical scale, not just intervals. To this end, he came up with a very simple process for generating the scale based on intervals, in fact, using just two intervals, the octave and the perfect fifth.

The method is as follows: we start on any note, in this example we will use D. This is the first note of the scale. If we go up by an octave, we again reach a D, one octave higher. This defines the range of the scale that we want to create. We want to fill in the notes of the scale between these two D's.

To get the next note, we start on the original D and go up by a perfect fifth. This gives us an A. We can also go down by a perfect fifth that takes us to a G below the original D. Although this is a new note, it is not in the range that we specified above. So, we simply raise it by an octave, which does not change the musical characteristic of the note. So far, we have D, G, A, D. Notice that these are just the notes we discussed in Section 2.2, Table 2.1, but now we can keep going.

From A, we can again go up by a perfect fifth, which takes us to E. However, this note is again outside the original scale that we are creating, so we bring the note down by an octave. Similarly, from G we go down by a perfect fifth to C and bring it up by an octave. The notes that we now have are D, E, G, A, C, D. If we play these notes on a keyboard, you may recognize them as a pentatonic scale. This scale is used in many musical styles, including Asian and Jazz.

Perhaps you can now see the process. Going up and down by perfect fifths generates new notes. However, sometimes these notes are outside the original scale that we are trying to create. Whenever a new note is outside this range, we move it up or down by an octave to put it in range.

You may not be familiar with the names of the notes, and this is not so important. What is very important is to know the frequency of each note with respect to the original note of the scale. So, we will review the process but now keeping track of the frequencies of each note. Since the ratios start to get complicated we will also convert them to simple decimal notation.

Remember, we are trying to fill the notes in within one octave. Since we only care about the relationships between the notes, we give the first note a frequency of 1 and the last note a frequency of 2:

Note	Ratio	Decimal
D	2	2.000
↑		
D	1	1.000

Now let's add notes by going up and down by a perfect fifth from the D: solid arrows are perfect fifths, dashed arrows are octaves.

Note	Ratio	Decimal
D	2	2.000
A	3/2	1.500
G	4/3	1.333
D	1	1.000
G	2/3	0.667

This gives the notes D, G, A, D, and the ratios 1, 4/3, 3/2, 2. If we continue to go up and down by fifths and move the notes by octaves, when needed, we get:

Note	Ratio	Decimal
E	9/4	2.250
D	2	2.000
C	16/9	1.778
A	3/2	1.500
G	4/3	1.333
E	9/8	1.125
D	1	1.000
C		0.889

This, again, is the pentatonic scale, D, E, G, A, C, D. If we do this once more, we get the Western scale, also called the diatonic scale.

Note	Ratio	Decimal
D	2	2.000
C	16/9	1.778
B	27/16	1.688
A	3/2	1.500
G	4/3	1.333
F	32/27	1.185
E	9/8	1.125
D	1	1.000

This is the Western scale starting on D. However, we could have started on any note and created the scale in the same way. More importantly, there are seven different scales that we could create in this way—essentially starting on any white key on a piano and playing a scale, staying on the white keys. Many of the scales sound unfamiliar to Western classical music, but they are in use in various musical systems. This came about because of the way we went up and down by fifths to generate the scale we found above. In that case, we went up by fifths three times to generate three new notes, and down by fifths three times, to generate three more notes. This gives six notes, plus the original one, yielding a seven-note scale. But, we could have kept going up by fifths six times, or down by fifths six times. In each case, we would produce a seven-note scale, each having a distinctive sound. In Western music, the scale starting on C is the common major scale (called the Ionian scale), while the one starting on A is the minor scale (the Aeolian scale). The major and minor scales are used in the vast majority of classical works. The scale we found above is called the Dorian scale and was sometimes used by Brahms, among others.

Circle of Fifths

What happens if we keep on going? This way of adding notes by going up and down by perfect fifths can be organized in a diagram called the Circle of Fifths (Fig. 2.6).

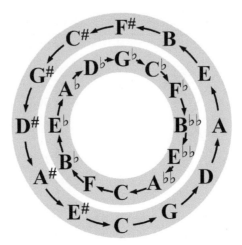

Figure 2.6 The circle of fifths. The # sign raises the note by a half step and the ♭ sign lowers the note by a half step.

This circle shows what note you arrive at by going up or down a fifth from any other note. As you can see, the inner or outer circle has more than the seven notes that we have already generated. This means that we stopped too soon. If we repeat the process a few more times, we arrive at the chromatic scale. This is the scale you hear if you play every note on the piano, white notes and black notes. There are a total of 12 notes in the chromatic scale, as shown in Fig. 2.6.

And what happens if we keep adding more notes in the same way? Do we start repeating? The answer is "yes" and "no". The question is whether the inner circle in the Circle of Fifths is the same as the outer circle. In other words, is a D-sharp (D#) the same as an E-flat (E♭)? On a piano, they are the same, but the exact frequency that you arrive at using the Pythagorean system gives different values for these two notes. So, they are not exactly the same. In fact, the Arabic scale considers these notes to be different. This is not a scale that we are familiar with in Western music. It contains notes in between our smallest division and has many more notes than we are used to, about 24 notes in the scale. However, it is perfectly normal in Arabic music.

Does this process ever stop? Can we keep adding notes indefinitely? The answer to the latter question is "yes", we can always keep generating new notes in this way. For the process to stop, it would have to start repeating at some point. In other words, if we eventually got back to D, we would then simply start over again and not produce any new notes. However, this will never happen. The reasoning is a bit tricky: say we always go up by perfect fifths. If we go up n times, the frequency is $(3/2)^n$. Now we go down by octaves m times. This takes us to $(1/2)^m$. Our final frequency is $(3/2)^n(1/2)^m$. Can this ratio ever equal 1? Can we ever get back to the starting note? For $(3/2)^n(1/2)^m = 1$, it would have to be true that $(3/2)^n = 2^m$. Rearranging this, we ask can $3^n = 2^{m-n}$? However, 2^{m-n} is *always* an even number, while 3^n is *always* an odd number. Therefore, it is *never* true that $(3/2)^n = 2^m$. Thus, we always keep adding new notes. Eventually, the notes get so close together that we will not be able to hear the difference, but we will return to this question later in the course.

Returning to the Western scale, there is another important feature of a scale, and that is the interval between adjacent notes in the scale. Here, again, is our final Western scale, based on the Pythagorean system, including the interval between each note, Table 2.2:

Table 2.2 The Dorian diatonic scale in Pythagorean tuning

Note	Decimal	Ratio	Interval between notes
D	2.000	2	9/8
C	1.778	16/9	256/243
B	1.688	27/16	9/8
A	1.500	3/2	9/8
G	1.333	4/3	9/8
F	1.185	32/27	256/243
E	1.125	9/8	9/8
D	1.000	1	

The ratios of the individual notes are getting complicated and the intervals between the notes are getting quite complicated, but, amazingly, there are only two such intervals, 9/8 and 256/243. We can add these new intervals to our list. They are called the whole step and the half step, respectively:

Whole step = 9/8

Half step = 256/243

This is a good check that the system is working. The idea of a scale is that there is some regular pattern to the sequence of notes in a scale. You may have noticed that the intervals between the notes of a scale are not all equal—that would lead to a rather uninteresting scale. But, they are not all different, either, as that would be very confusing. As it turns out, the Pythagorean scale is made up of whole steps and half steps in a particular pattern. There is one small difference from what we might have expected. From their names, you might think that two half steps are equal to one whole step. Is this true?

2.5   Space-Time Connection

Waves in Space

So far, we have focused on the fact that waves repeat in time. The time it takes to repeat is called the period and the number of times it repeats in 1 s is called the frequency. However, if you ever sat at the ocean and watched water waves roll in, you would realize that waves also repeat in space. So, a better definition of a wave is something that repeats in time *and* in space. This introduces a new important quantity: how far does it take for a wave to repeat in space? This distance is called a wavelength:

A wave repeats in time in one period. A wave repeats in space in one *wavelength*.

For example, again imagine you are at the beach watching the waves in the ocean. The time between when the waves break is the period of the wave and the distance between crests while the waves are out in the ocean is the wavelength. Physicists often associate Greek letters with certain quantities and in this case, the wavelength is usually given the Greek letter for "*l*" or λ.

Waves on a string, then, will have a wavelength, as well. Look back at Fig. 2.5. Those are all examples of waves in space on a string. What you should notice is that a wavelength is twice the distance between nodes. After one node, it looks like the wave is repeating, except that it is upside down. Only after two nodes is the wave really repeating.

Worked Example

I have a 1 meter length of string which vibrates at f_a = 500 Hz in its fundamental mode.

a. What is the wavelength of this mode, λ_a?

b. I place a bridge 1/3 of the way from one end, so that only the long portion can vibrate. What are the frequency and the wavelength of the new fundamental, f_b, λ_b?

c. Now, I only lightly touch the string when I pluck it. What are the frequency and the wavelength now, f_c, λ_c?

d. What is the interval between f_b and f_c?

Solution

a. From Fig. 2.5, we see that, in the fundamental mode, there are only two nodes, which corresponds to half of a wavelength. So, the full wavelength is twice the length, or λ_a = 2 m.

b. Now, the length of the string is 2/3 as long, so the frequency is 3/2 higher, or f_b = (3/2)(500 Hz) = 750 Hz. Also, the string is now (2/3)(1 m) = 0.667 m long, so the wavelength is again twice this, or λ_b = 1.334 m.

c. Again, from Fig. 2.5, we see that if we lightly touch the string, we place a node at the 1/3 point and the whole string still vibrates, producing a frequency of f_c = $3f_a$ = 1,500 Hz. Similarly, the wavelength is one third as long, or λ_c = (2 m)/3 = 0.667 m.

d. f_c/f_b = (1,500 Hz)/(750 Hz) = 2, which corresponds to an internal of an octave.

Space-Time Connection

Often we interchange space and time. If someone asks you how far it is to Boston, you might say 90 miles or an hour and a half. Both answers would be regarded as perfectly normal. What is needed to make this connection? *A speed or a velocity.* In traveling by car, it is easy to assume the speed is 60 miles/h or 1 mile/min. Then the conversion is 1 mile corresponds to 1 min of travel.

To make the connection between the frequency and the wavelength of a wave, we also need a velocity: *the velocity of the wave.* What is the velocity for sound waves? It is very fast, about 343 m/s, or about 767 miles/h. The concept of a wave velocity simplifies things considerably, because it connects frequency and wavelength in a very simple way:

$$\text{Frequency} \cdot \text{wavelength} = \text{Velocity, or } f \cdot \lambda = v \qquad (2.1)$$

Wavelength is a distance, so it is usually given in meters. As we know, f is the frequency and has units of Hz = 1/s. So the product $f \cdot \lambda$ has units of m/s, which, in fact, is a velocity. The symbol for velocity is "v". Equation 2.1 is the most important relationship in the course—you will use it many times. You can also verify this relationship in Lab 2.2.

How does all of this relate to a vibrating string? Remember, the frequency of a string is given by Eq. 1.2. However, we now have to amend this formula, as we found, in this chapter, that strings can vibrate in more than one mode. We specify the mode with a number n. And, we found that the frequency of the nth mode is nf, where f is the fundamental frequency. So, Eq. 1.2 applies to the fundamental, but we can now write, more generally, that the frequency for the nth mode is:

$$f = \frac{n}{2L}\sqrt{\frac{T}{U}}. \qquad (2.2)$$

The distance between nodes is L/n so the wavelength is $\lambda = 2L/n$. Let's rewrite the equation a little differently:

$$f\frac{2L}{n} = \sqrt{\frac{T}{U}}, \qquad (2.3)$$

so we can write $f \cdot \lambda = \sqrt{\frac{T}{U}}$. Thus, we see that:

$$v = \sqrt{\frac{T}{U}}. \qquad (2.4)$$

This is the velocity of a wave on a string with tension T and density U. For different waves, we just need to know or measure the velocity of the wave. Then you can convert from wavelength to frequency and back using Eq. 2.1.

The velocity depends on the details of the particular kind of wave. For example, the velocity of a wave on a string depends on the tension and density of the string. As we will see, the velocity of sound depends on the pressure and density of air.

The wavelength does not depend on the details of a vibrating system: it does not depend on the tension or the density for waves on a string. However, it does depend on the *geometry*: the wavelength on a string does depend on the length of the string and the number of nodes. Finally, we know that the distance between nodes is half a wavelength. So is the distance between antinodes. This means that the distance between a node and an antinode is a quarter of a wavelength, or $\lambda/4$.

2.6 Air Columns and Boundary Conditions

Vibrating strings are important to study in the physics of music, as all string instruments are based on properties of the vibrating string. However, there are many other musical instruments that do not involve strings. These are the woodwind and brass instruments. To understand these instruments, we must consider a new type of wave: waves in air. These waves also give rise to sound waves, which are the waves that allow us to hear anything, in the first place.

When discussing waves by using pictures, we must decide how to represent the wave. For the string this was easy. Because we can see the wave, it makes sense to just plot the displacement of the string. But what about a sound wave? We cannot see the wave directly, so its representation is a little more abstract. Sound waves are created when there is a change in pressure. Consider what happens when you make the sound of a "p". The air pressure in your mouth increases until you suddenly release the air making a puffing sound. This disturbance in the air pressure then moves as a wave. So, when we discuss sound waves, we will be talking about the pressure change of the air due to a wave.

Let's start by considering a narrow tube (like a flute) that is closed at both ends. Initially, the pressure in the tube is constant (Fig. 2.7).

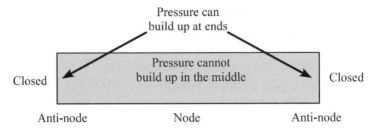

Figure 2.7 Closed-closed air column.

The air in the tube can move along the tube. At the ends of the tube, the air piles up and can increase the air pressure at that point. In fact, it is at the end of the tube that the pressure increase can be the maximum. So, for a wave in a tube closed at both ends, there will be antinodes at the ends of the tube. Notice that this is just the opposite of vibrating strings. In the middle of the tube, the air moves back and forth but the air pressure does not change—there is nothing for the air to push or pull against to increase or decrease the air pressure. So, there is a node in the middle of the tube (at least for the fundamental mode of vibration). If the air sloshes towards one end of the tube the pressure will increase at that end, but there will now be too few air molecules at the other end of the tube and so the pressure will decrease. However, the high pressure at one end will then force the air back to the other end, reversing the situation—this is one half of an oscillation of the air in the tube.

Using the fact that the closed tube has antinodes at each end of the tube, we can make plots of the first few modes of vibration (Fig. 2.8).

If the length of the tube is L, the wavelengths of these modes are: $2L$, L, and $2L/3$. If we also know the speed of sound, we can determine the frequency of these modes from $f = v/\lambda$, or $v/(2L)$, v/L, $3v/(2L)$. In general, the frequencies of closed-closed air column in the nth mode are, $f_{\text{c-c}}$:

$$f_{\text{c-c}} = n\frac{v}{2L} \tag{2.5}$$

This produces a harmonic overtone series which we can demonstrate with Lab 2.3.

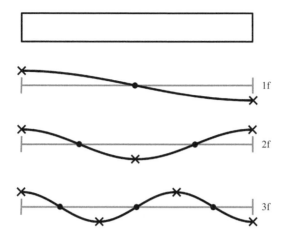

Figure 2.8 Vibrational modes of a closed-closed air column.

Now, an air column has a new feature compared to strings: we can have a different kind of end, an open end. If a closed end creates a pressure antinode, an open end will have a pressure node (Fig. 2.9).

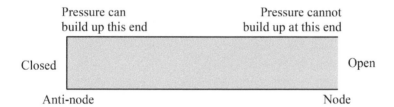

Figure 2.9 Open-closed air column.

What do the lowest modes of vibration look like in this case of one closed end and one open end? Just remember that there must be an antinode at the closed end and a node at the open end (Fig. 2.10).

Since the distance between a node and antinode is a quarter of a wavelength, $\lambda/4$, the wavelength of the fundamental mode will be $\lambda = 4L$. The wavelengths of the next two modes are $4L/3$ and $4L/5$. The frequencies of these modes will then be $v/(4L)$, $3v/(4L)$, and $5v/(4L)$. In this case, the modes of the open-closed air column in the nth mode, $f_{\text{o-c}}$ are:

$$f_{o-c} = (2n+1)\frac{v}{4L} \tag{2.6}$$

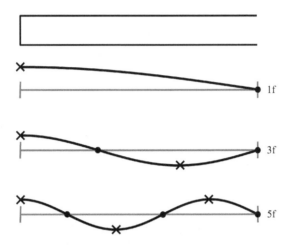

Figure 2.10 Vibrational modes of an open-closed air column.

In general, whether you have a node or an antinode at one end of a string or tube or any vibrating system is called a *boundary condition.* The boundary conditions for an air column are:

- Closed end: antinode.
- Open end: node.

These boundary conditions determine the possible modes of vibration. From the modes of vibration, we can determine the wavelength of each mode. Once we have the wavelength, we can find the vibrational frequency from the expression: $f \cdot \lambda = v$.

As it turns out, different boundary conditions give different overtone series:

- A closed-closed tube has a series: *1f, 2f, 3f*, ..., which is a harmonic series.
- A closed-open tube has a series: *1f, 3f, 5f*, ..., which is a partial harmonic series.

What about an open-open tube? In this case, there must be a node at each end. If we sketch the possible modes, they will look just like the modes of a vibrating string and will have the same overtone series (see Fig. 2.5). So, an open-open tube has a harmonic overtone series: 1*f*, 2*f*, 3*f*, ...

What effect do these different possibilities have on musical instruments? From the three cases above (c-c, c-o, o-o) we see that there are two possible overtone series: 1*f*, 2*f*, 3*f*, ... and 1*f* 3*f*, 5*f*, ... As it turns out, almost all woodwind and brass instruments have a full harmonic series. The one exception is the clarinet; it has the partial harmonic series. This is what gives the clarinet a very distinctive tone—it is more "open" and "pure" because of the fewer harmonics in its tone. But, we will come back to the relationship between tone quality and overtone series, later.

Now, just because the c-c and o-o tube have the same overtone series does not mean that they are identical—their modes of vibration, i.e., the placement of the nodes and antinodes, are quite different and this can affect the placement of finger holes in a wind instrument. Remember that there are two ways to change the pitch of a string: we can press down on a string to shorten the length that vibrates. This raises the pitch of the string. Or, we can lightly touch the string with our finger. This allows the entire string to vibrate, but it imposes a node at the point where we are touching the string. If there is a mode of vibration with a node at that particular point, we will hear that mode, which will have a higher pitch than the fundamental. This technique is used frequently by string players to produce notes with a very high pitch.

The same two ways exist to change the pitch of a wind instrument. First, the length of the tube determines the pitch of the fundamental. Although the length of a flute is always the same, the effective length is determined by how many of the finger holes are covered. The more that are covered, the lower the pitch. However, just as with strings, we can produce higher harmonics on a wind instrument. The effect of an open finger hole is the same as the open boundary condition for a tube: it imposes a node at that point. So, in addition to the regular holes which effectively change the length of the instrument, wind instruments have register holes which place nodes along the length of the instrument. If there is a vibrational mode with a node at the open register hole, we will hear that particular mode, with its particular pitch. Lab 2.4 explores the effects of a register hole in a closed-closed, open-closed, and open-open tube (the latter discussed below).

A flute acts like an open-open tube. Although one end is closed, the part of the flute between the embouchure hole (where the flutist blows into the instrument) and the far, open end, acts like an open-open tube. This gives the flute a full harmonic overtone series. To produce the second mode of vibration, a register hole could be placed at the halfway point (see Section 2.3). This hole would suppress the fundamental and favor the second harmonic. In this way, a flutist could play a key, which opens such a hole, and raise the pitch of a note by a factor of 2, or an octave. Real flutes actually don't need such a register key, as a good flutist can reach the second register by blowing in different ways.

For another example, consider a closed-open tube in its fundamental mode, shown above. If we place a hole halfway along the tube, we can see from the modes in Fig. 2.10 that none will vibrate, as no mode of a closed-open tube has a node in the middle of the tube. But what happens now if we open a hole one third of the way from the closed end of the tube? The air will vibrate if there is a mode of vibration with a node at this same position. As it turns out, there is such a mode (Fig. 2.11).

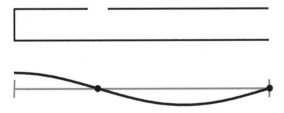

Figure 2.11 Effect of a register hole on a closed-open tube.

In the first case, the mode had a frequency of f, while in the second case, the frequency is raised up to $3f$. Rather than raising the pitch by an octave, as in the case of a flute, the pitch now goes up by an octave and a fifth ($2 \times 3/2 = 3$). Interestingly, the clarinet acts like a closed-open tube, as the end with the reed is essentially closed. And, indeed, the clarinet actually does use such a register hole to obtain a higher register, but, now, this register is a full octave and a fifth higher. This complicates the playing of a clarinet, compared to a flute. On a flute, a note has a particular set of finger holes that need to be open or closed, called the fingering which is the same for the first and second register. In contrast, on a clarinet, if one plays a particular note, say a "C", and opens the register hole, the pitch will go up to a "G" an octave higher. Thus, the same fingering on a clarinet produces different notes in the two registers. Thus, the clarinetist has to learn two sets of fingering, one for each register. Incidentally, the rest of the woodwind instruments (oboes, bassoons, saxophones, etc.), all behave like the flute—an open-open tube with the upper register an octave higher. (These instruments might appear to be open-closed, as they all have reeds at one end. However, they are not straight tubes—they are actually long cones. The diameter of the tube increases from the reed all the way to the end. These types of tubes are harder to analyze, but end up having a full harmonic overtones series.) Finally, brass instruments also have a harmonic overtone series and but they play on many different registers, which is achieved in a different manner.

Vibrating strings and air columns form the basis of all melodic instruments in an orchestra and the physics of these two systems determine how these instruments sound and are played. We will discuss the other major class of instruments in the next chapter and explore what makes the fundamental difference between melodic and percussion instruments.

End Correction

Before we leave the topic of air columns, we must consider two very significant corrections to the resonant frequency of an air column—one that has a large effect on the construction of wind instruments. This effect is called the "end correction" and it concerns the open end of a tube. Above, we stated that there is a pressure node at the open end of an air column. This is essentially correct, except that the node is not precisely at the end of the tube; it is a little way outside the tube (Fig. 2.12).

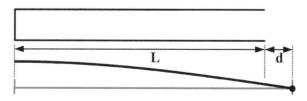

Figure 2.12 End correction to an open end of an air column.

The question is how much beyond the end of the tube does the node occur? Here, one needs calculus to figure this out, but scientists have found that the extra distance, d, is related to the *radius* of the tube in the following way:

$$d = 0.6\,R, \tag{2.7}$$

where R is the radius of the tube.

So, a very thin tube has almost no end correction, while a fat tube will have a large end correction. How does the end correction affect the resonant frequencies of the tube? It is fairly straightforward to take this into account. We generally say that the tube has a length L, for finding the wavelengths of the different modes. However, it is not the length that really matters, it is the distance between nodes and antinodes. For a closed-open tube, the antinode sits right at the closed end and the node sits at a distance $L + d$ away. So, we can simply say that the tube has an *effective length* $L_{eff} = L + d = L + 0.6\,R$. We use the term effective length, because the tube acts as though that is its length, even if its physical length is still L. Notice that if we have an open-open tube, we have to apply the end correction twice: $L_{eff} = L + 2(0.6)R$. The end correction can be measured with Lab 2.5.

Worked Example

Consider an open-open tube with a length of 45 cm and a diameter of 4 cm. What is the fundamental frequency of the tube with and without the end correction? By what interval would including the end correction change the pitch?

Without the end correction, the wavelength of the fundamental would be $\lambda = 2L = 90$ cm. This would give a fundamental frequency of $v/\lambda = (343 \text{ m/s})/(0.900 \text{ m}) = 381$ Hz.

With the end correction the effective length will be $L_{eff} = L + 2(0.6R) = 45$ cm $+ 2(0.6(2 \text{ cm})) = 47.4$ cm. Remember, the radius of this tube is 2 cm and there are two open ends. So now, the wavelength of the fundamental is $\lambda = 2 \times 47.4$ cm $= 94.8$ cm and the frequency of the fundamental is $(343 \text{ m/s})/(0.948 \text{ m}) = 362$ Hz.

The ratio of the frequencies is $(381 \text{ Hz})/(362 \text{ Hz}) = 1.052$, which is almost a Pythagorean half step $(256/243)$. So, the end correction in this example lowers the pitch by a half step, which, musically, is a large effect.

Dependence of the Speed of Sound on Temperature

The other correction is that the speed of sound changes with temperature. Around room temperature, the dependence of sound speed on temperature is:

$$v = (331.3 + 0.6T)\,\text{m/s}, \tag{2.8}$$

where T is the temperature in °C. This is important when, for example, a marching band tunes their instruments in a cool gym and then goes out onto a hot football field. The change in temperature can be significant, causing the tuning of the wind and brass instruments to change.

2.7 Music of the Spheres and the Lessons of Pythagoras

Conclusions So Far

At this point, we have covered a number of topics and some history of physics, so it is important to step back and understand where we are before going on.

I. Using simple mathematics, Pythagoras was able to describe the basis of almost all musical scales, including the pentatonic, the Western, the chromatic, and the Arabic scales. This shows the power and excitement of science. For the first time, Pythagoras could answer the question, *why*? Why are these notes and scales special? The answer is that they are formed in a simple, systematic, and mathematical manner. Most importantly, Pythagoras showed that *the notes are not random or arbitrary and that they could be understood on a deeper level.*

II. Pythagoras' discoveries bring up a deeper 'psychology' question: scales were first developed by ear: we—and perhaps the Neanderthals—chose these particular notes before there was any understanding of mathematics or physics. The notes were chosen simply because they were pleasing to the ear. But, as it turns out, the scales also follow basic mathematical constructs. So the question is, what does this say about our likes and emotions? Is there, at least partially, a mathematical/physical basis to them?

III. The power of spectroscopy. What Pythagoras did was look at a physical system (the musical scale), found characteristic frequencies (pitches/notes) and found simple mathematical relationships between the frequencies (ratios of 3/2, for example). This process actually became a fundamental part of physics, and modern physics, in particular.

Modern physics started in the early 1900s with the concepts of quantum mechanics and relativity. However, the concept of quantum mechanics came about through a process much like Pythagoras'. A physicist, Johann Balmer, considered a physical system (the hydrogen atom), examined the frequencies of the colors of light that a hydrogen lamp produced, and found simple relationships between the frequencies (such as ratios of $3^2/4^2$). Another physicist, Neils Bohr, built on Balmer's observation and proposed a radically new structure of the hydrogen atom, which sparked the development of quantum mechanics.

This process, used by Pythagoras, Balmer, Bohr, and many others, is called spectroscopy (loosely, the study of frequencies) and is at the heart of most modern science. To better understand it, spectroscopy also forms the basis of most of the labs in the Appendix. Indeed, strings and air columns fall into the same process:

Physical system	Musical scale	Hydrogen atom	Air column
Frequencies	Notes of scale	Colors of light	Resonances
Relationship	n/m	n^2/m^2	$n \cdot f$

Many Caveats to This Scientific Success

Although Pythagoras ushered in a whole new way of looking at the world, there was still much more to be done in the development of modern science. So, it is also important to understand the limitations up to this point.

I. *Pythagoras got lucky*: Pythagoras did not actually study the frequencies that made up pleasing intervals and the musical scale. Rather, he made observations about the lengths of the strings that made intervals and scales. However, it just so happens that the frequency of a string is simply related (i.e., inversely proportional) to the length of the string.

So, all the conclusions he reached about ratios of lengths of strings for different intervals also apply to the ratios of the frequencies in the intervals.

The problem is that this is not true for the tension in the string, for example. To study how frequency depends on the parameters of the string, you could use an apparatus like this where you can change the tension on the string by hanging different weights from the string (Lab 2.1) (Fig. 2.13).

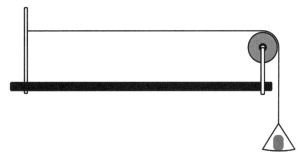

Figure 2.13 Apparatus to determine the dependence of frequency on tension, length, and density.

Pythagoras might have tried to use an apparatus like this to find relationships between the weights needed to form pleasing intervals. However, if you double the weight, the frequency changes by $\sqrt{2}$ not 2; in other words, the ratios for the intervals would not be the same as for the ratio of the weights. Pythagoras might not have achieved his goal of understanding the musical scale if he used weights instead of lengths.

Also, the frequency is not simply related to the length when it comes to vibrating bars, like you find on a xylophone or a marimba. As we will see in Chapter 3, the frequency is inversely proportional to the *square* of the length. If Pythagoras had studied a xylophone, instead of a lyre, things, again, might have turned out quite differently.

Although these possibilities may seem a bit far-fetched, they are not, as can be seen from a famous woodcut made in the Middle Ages to explain the theories of Pythagoras (Fig. 2.14).

Figure 2.14 Medieval woodcut.

The purpose of the woodcut was to illustrate the Pythagorean ratios and how they applied to musical instruments. The lower right panel shows flutes whose lengths correspond to the Pythagorean ratios. Notice how the ratios of any of the numbers reduce to a Pythagorean interval, for example 16/12 = 4/3, a Perfect Fourth. This works out just fine because, like a string, the frequency of an air column (e.g., a flute) is simply related to its length. However, the panel on the lower left has a problem. The weights on the ends of the strings will change the tension in the strings. But, as discussed above, the frequency of the string is not simply related to the tension, and the strings will not sound according to the Pythagorean intervals. The examples in the upper panels are even more complicated, but suffice it to say, the bells, water glasses, and anvils will not produce the correct intervals.

One of the defining features of the Middle Ages was a reverence for the knowledge passed down from the ancient Greek philosophers. This knowledge was then applied to situations where it did not apply, but there was also no tradition of actually trying something out to see if it was correct. Had anyone tried to build any of the instruments (except for the flute) in the ratios prescribed in the woodcut, they would have found that the intervals were all wrong.

A crucial step for the development of modern science was the willingness to test the theories of others to see if they really were correct or not. In fact, it was Vincenzo Galilei, Galileo Galilei's father, who first noticed that there was something wrong with the woodcut. This period marked the beginning of the Renaissance, and we will hear more about Vincenzo later. However, he certainly set the stage for his son to come along and challenge the most basic beliefs about the Universe at the time.

II. While Pythagoras was making a lot of progress in mathematics, geometry and music, the Greek astronomers of the time were not doing quite so well. They realized that there were certain "fixed" stars—stars whose relative position in the sky did not change through the seasons. They also noted that there were "wanderers" or planets (planet is the Greek word for wanderer). These planets moved around relative to the background stars.

To explain these observations, the astronomers figured that the fixed stars were attached to a large black sphere that defined the edge of the Universe. The planets had to be attached to moving spheres, with each planet on its own sphere. However, these spheres could not be black, otherwise, one could not see through to the stars in back. So, the spheres that the planets were attached to had to be crystal spheres. The "planets" that the astronomers knew about were: the moon, Mercury, Venus, the Sun, Mars, Jupiter, and Saturn. Seven, in all, so there were seven crystal spheres. The big question was: why seven? To the Greeks, the answer was obvious: Pythagoras just had his big breakthrough that mathematics could explain phenomena in nature and he now understood why there were seven notes in the musical scale. Thus, there must be the same reason for why there are seven crystal spheres. The astronomers were so convinced of this that they called it the Music of the Spheres. This concept was so powerful that it led astronomy astray for 1500 years!

Figure 2.15 Music of the Spheres.

Even many great physicists were taken in by this vision that the Universe had an order that was musical in nature. The astronomer Kepler became obsessed with trying to fit the orbits of the planets to a musical scale. This, again, demonstrates the danger of taking knowledge from one field and applying to other fields where it is not appropriate or applicable.

The great physicist Sir Isaac Newton was also drawn in by Pythagoras' success, even attributing his greatest discovery to Pythagoras:

[I]*n* general terms, if two strings equal in thickness are stretched by weights appended, these strings will be in unison when the weights are reciprocally as the squares of the lengths of the strings. Now, this argument is subtle, yet became known to the ancients. [Pythagoras] applied [this] to the heavens and consequently by comparing those weights with the weights of the Planets and the lengths of the strings with the distances of the Planets, he understood by means of the harmony of the heavens that the weights of the Planets towards the Sun were reciprocally as the square of their distances from the Sun.—*Isaac Newton*

However, as an inspirational force, music did aid in the development of physics long after the planetary orbits were understood. Most of the founders of quantum mechanics started out by studying music. Strangely enough, the motion of the electron in a hydrogen atom actually does follow the Music of the Spheres, to a certain extent.

III. Just because Pythagoras found a very simple way of constructing the musical scale, it does not mean that it is the *only* way, or even the simplest way. In fact, there are many different ways of constructing the scale and the Pythagorean way is only used infrequently.

For example, a piano is tuned on the basis that two half steps should equal one whole step. In the Pythagorean scale a half step = 256/243. So, two half steps give $(256/243)^2$ = 1.110, while the whole step is 9/8 = 1.125. These are close, but you can hear the difference. On a piano, a whole step = 1.122, in between the two.

So, just because Pythagoras found a simple mathematical answer, it may not be the whole answer or even the only answer.

IV. There are many things that Pythagoras did not explain. Why is the frequency inversely proportional to the length? How does tension affect the pitch? Do different types of strings affect the pitch? *Why* do we like these intervals? However, Pythagoras did make a great start.

2.8 New Scales and the Development of Western Harmony

Even though Pythagoras came up with a simple mathematical way of understanding the musical scale, not everyone agreed with his construction. In particular, the philosopher, Ptolemy, who lived around 140 A.D., was not happy with the Pythagorean scale. He felt that the intervals became too complicated.

Looking back at Table 2.2 showing the Pythagorean scale, one cannot really blame Ptolemy for saying that it was not so simple, after all. As we saw before, there are only two steps in the scale: the whole step = 9/8 and the half step = 256/243. However, the ratio for the half step is not simple and, even worse, two half steps do not equal one whole step. There is another subtle problem concerning the perfect fifths. Pythagoras based the whole concept of building a scale on the perfect fifth and indeed, within the scale in Table 2.2, all fifths are perfect. However, if we continue the process and generate a chromatic scale, eventually one interval of a fifth must be imperfect. This can be seen in Fig. 2.6—if we go all the way around the circle in either direction, we don't come back to where we started. That is why the process goes on forever. But, this also means the last step, $E^{\#}$ to C, cannot be a perfect fifth.

While the imperfections above are somewhat subtle, there are other problems which create more immediate problems for musical harmony, for example the major third. The major third should be a pleasing musical interval corresponding to F to A, or G to B on the piano but in the Pythagorean scale it corresponds to a rather awkward ratio of $(81/64) = 1.266$. In fact, besides not having a simple ratio, the Pythagorean major third does not sound good. (We will see why in Chapter 6.) As a result, Pythagoras' solution to this problem was to simply ignore the major third. Because of this, the Church even banned the use of the major third during the Middle Ages:

> In 1324, Pope John XXII decried that only "... concords such as the octave, fifth and fourth, that enrich the melody and may be sung above the simple ecclesiastical chant," will be permitted.

This inhibited the development of Western music and harmony for a long time. At that time, Gregorian chant was developed, as it does not use the major third. Music without the major third sounds quite stark and open, characteristics of Gregorian chant. Again, it was Vincenzo Galilei, among others, who pushed for a change in the definition of the major third that would allow for more complex harmony.

The Just Scale

Back to Ptolemy: Ptolemy had a different idea of simplicity. He wanted all basic intervals to be simple, not only the fourth, fifth, and octave. These are known as the Just intervals. So, instead of 81/64 for the major third, he chose the closest simple ratio, which is 5/4 = 1.250. Remember, Pythagoras formed all his basic intervals with the numbers 1, 2, 3, and 4 and he did not want to go on to use 5. However, it turns out that 5/4 is really the natural way to define the major third.

Ptolemy also noticed that starting from the first note in the scale, going up a third and then a perfect fifth gives the frequencies, $1f$, $(5/4)f$, $(3/2)f$. Or, to put it a different way, this group of three notes are in the ratios of 4:5:6. In fact, this group of notes has a special name: the major triad. The major triad is perhaps the most familiar and important chord in Western music as it establishes the tonic or starting note of the scale and the key that the piece is written in. Here, key means the sharps or flats that are used throughout a piece of music.

Interestingly, this sequence of 4, 5, and 6 is part of a harmonic series. So, if one plays a note with a complete harmonic series, it contains the frequencies, $1f$, $2f$, $3f$, $4f$, $5f$, $6f$, ... With this definition of the major triad, we see that *any* note with a harmonic overtone series has contained within its overtones a major triad. This is why the major triad is so important musically. Moreover, this is also why it is so important to define the major third as 5/4. Otherwise, it would not fit into the harmonic series. The major triad is warm, rich, and pleasing as long as it uses the Just major third. If played with the Pythagorean major third, it sounds quite bad, as discussed above, which is why the Pythagorean interval was rejected. We will return to this question of overtone series, tone quality and consonance versus dissonance in detail, in Chapter 6 and Lab 6.2 will demonstrate the problem with the Pythagorean major triad.

So, Ptolemy had good esthetic and musical reasons for wanting a different definition for the major third. But, how did he construct a musical scale? His goal was to preserve the harmony of the major triad, as this was the most important reason for introducing the new definition. To demonstrate his construction, let's start on a C and build a major triad. This gives us two new notes, E and G:

Note	Ratio	Decimal
C	2	2.000
G	3/2	1.500
E	5/4	1.250
C	1	1.000

Just as Pythagoras built on the previous intervals, so did Ptolemy, meaning that he then added a major triad starting on G, with the same ratios. Also, if a note got too high or low, it would be adjusted by octaves, again, just like Pythagoras:

Note		Ratio	Decimal
	D	9/4	
C		2	2.000
	B	15/8	1.875
	G	3/2	1.500
E		5/4	1.250
	D	9/8	1.125
C		1	1.000

Finally, Ptolemy needed one more triad, which he arranged to *end* on C:

Note	Ratio	Decimal
C	2	2.000
B	15/8	1.875
	5/3	1.667
G	3/2	1.500
E	4/3	1.333
D	5/4	1.250
	9/8	1.125
	1	1.000
	5/6	
	2/3	

The final Just scale, including the individual steps in the scale, is shown in Table 2.3:

Table 2.3 C major scale in Just tuning

Note	Decimal	Ratio	Interval between notes
C	2.000	2	16/15
B	1.875	15/8	9/8
A	1.667	5/3	10/9
G	1.500	3/2	9/8
F	1.333	4/3	16/15
E	1.250	5/4	10/9
D	1.125	9/8	9/8
C	1.000	1	

Notice how the Just intervals are much simpler than the Pythagorean intervals. This is what Ptolemy was after. Especially nice is the fact that all major thirds are simple and the same. However, are there any new problems? The answer is emphatically, yes!

There are now *two* kinds of whole steps, 10/9 (major) and 9/8 (minor). Moreover, the half step is 16/15 and two half steps don't produce either kind of whole step. Having two kinds of whole steps is a particularly bad characteristic for a scale, because then a keyboard, like a piano, can only be tuned to one key at a time. If you want to play a second piece in a different key, you must retune the entire instrument.

However, there is an even bigger problem: not all of the fifths are perfect. The interval from D to A, which should be a perfect fifth comes out to $(5/3)/(9/8) = 40/27 = 1.481$ instead of $3/2 = 1.500$. This is very bad, indeed. After octaves, fifths are the most important interval, and it is very jarring to have them out of tune.

Is the Just scale better or worse than the Pythagorean scale? This is a difficult question to answer. Gregorian chant sounded fine with the Pythagorean scale, so if this is the music you like you don't need the Just scale. However, the Pythagorean scale will not allow you to develop the more complex harmonies of Western music.

The Just scale is based on three triads: the tonic, the dominant (the one built on the third note of the tonic) and the subdominant (the one whose third note is the tonic). Each of these triads has a particular role in Western harmony and the change from one to the next can be very pleasing and satisfying. These changes are called cadences and form the smallest unit in a musical composition.

As mentioned above, the biggest problem with the Just scale is that it is very hard to implement on a keyboard instrument like a harpsichord or a piano. The instrument had to be tuned differently for every key, which, of course, is very awkward. Harmonically, this is a big drawback, because it also means that any one piece must stay in the same key. So, even though more advanced harmony is possible, it could not be fully exploited because modulation, or movement between different keys, was not possible.

The Equal Temper Scale

Heroic attempts were made to construct keyboards which could play the Just scale in different keys. However, they had to have many extra keys. So, they were difficult to build, difficult to tune, and difficult to play. As a result, this solution did not last long. In the end, the Just scale was simply not adequate for the composers now trying to extend the scope of Western harmony. To fix the problems of both the Pythagorean and Just scales a more drastic solution was found. Rather than build up the scale from intervals or triads, the octave was simply divided up into 12 equal half steps. Furthermore, the whole step was chosen to be exactly two half steps. This represents yet a different definition of simplicity. This scale is called the Equal Temper scale, as each step is equal.

Now, for there to be exactly 12 half steps in an octave, it must be true that:

$$(\text{half step})^{12} = 2.$$

So the half step = 1.0595. Also, a whole step = $(\text{half step})^2$ = 1.123.

Since the Equal Temper scale is based on half steps, we need to see if the intervals come out correctly. A fifth is made up of seven half steps, so in the Equal Temper scale a fifth is given by $(\text{half step})^7 = 1.498$. This is not exactly 3/2, but it is very close. It is much better than the imperfect fifth in the Just scale. Moreover, since all half steps are the same, all fifths will be the same. Most importantly, in the Equal Temper scale, all keys are equal, even if no intervals are quite perfect. So, now a single piece could modulate from one key to another, as the keyboard did not have to be retuned for each key.

There are, in fact, quite a few solutions to creating musical scales and are called temperaments. The most common, besides the ones discussed above are Meantone Temperament and Well Temperament. Each tries to balance the pure Just intervals with the desire to have every key sound reasonable. By some measure, each of these makes the intervals more consonant than the Equal Temper scale, but, as a result, the different keys all sound slightly different, giving each of them subtly different characteristics or colors.

This breakthrough in tuning systems was captured by the great composer J. S. Bach when he wrote a Prelude and Fugue in every major and minor key—a pivotal work called the Well-Tempered Clavier, named after the tuning system used. It was the first collection of its kind with pieces in every key. It demonstrated that one could, indeed, now compose in all possible keys on a keyboard instrument—in Bach's case, a clavier.

Before we end this chapter, we must ask a very important question. We spent quite a bit of time discussing the different tunings, but we never checked to see if they make any noticeable difference. We argued that the major third in the Pythagorean tuning will sound out of tune. We also mentioned that the major triad is central to the Just tuning. So, the major triad highlights the differences between the different tuning systems. Lab 2.6 outlines how to generate a major triad in the three tuning systems that we have covered in this chapter and you can judge for yourself which major triad sounds the best and which sounds the worst.

2.9 Problems

Section 2.1

2.1 The Neanderthal bone is significant because it shows that Neanderthals probably (which one is not true):

(a) Used the Western musical scale.

(b) Were skilled craftsmen.

(c) Made musical instruments.

(d) Used bones for tools.

2.2 Pythagoras is important because (which one is not true):

(a) He studied the musical scale.

(b) He tried to explain phenomena with numbers.

(c) He invented calculus.

(d) He started a new method for analyzing the world around us.

2.3 A musical scale is:

(a) A group of notes that fit within an octave.

(b) The group of octaves that cover all the notes on a piano.

(c) The continuous range of frequencies within an octave.

(d) None of the above.

Section 2.2

2.4 If the ratio of the frequencies of two notes is 3, the higher note is:

(a) An octave higher.

(b) Two octaves higher.

(c) A perfect fifth lower.

(d) An octave and a perfect fifth higher.

2.5 If I go up an interval of a perfect fifth twice, what is the ratio of the new frequency to the old frequency?

(a) 25

(b) 2/3

(c) 9/4

(d) 3

2.6 Two strings have pitches of 630 and 840 Hz. What is the interval between them?

(a) Perfect fourth

(b) Unison

(c) Perfect fifth

(d) Octave

2.7 If we start at a frequency, f, go up by an interval of a fifth and then up by an interval of a fourth, what is our new frequency?

2.8 If we start at a frequency of 200 Hz, go up by an interval of a fifth and then down by an interval of a fourth, what is our new frequency?

2.9 (a) Starting from a frequency of 520 Hz, go up by three intervals of a fifth (3/2). What frequency are you at?

(b) Now bring this note down by octaves until it is in the octave starting on the note with a frequency of 520 Hz. What frequency are you at, now?

2.10 The lowest frequency on a standard piano is 27.5 Hz. If you are starting at a frequency of 250 Hz, what is the maximum number of perfect fifths you can go down by while still being in the piano range?

2.11 A natural trumpet (no valves) only plays on a harmonic series f, $2f$, $3f$, $4f$, $5f$, $6f$, $7f$, ...

How many notes are in the first octave? How many are there in the second?

What octave must the trumpeter work in to have at least seven notes?

Section 2.3

2.12 Which of the following is a proper mode of vibration for a vibrating string?

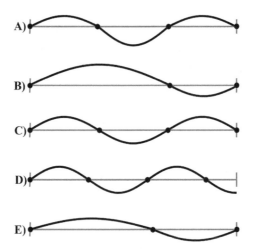

2.13 Match the questions with the answers:

How many nodes and antinodes does a string vibrating in its 3rd mode have?

How many nodes and antinodes does a string vibrating in its 4th mode have?

A string can never vibrate with this number of nodes and antinodes.

(a) Four nodes and three antinodes.

(b) Three nodes and four antinodes.

(c) Five nodes and four antinodes.

(d) Three nodes and two antinodes.

2.14 If f is the fundamental frequency of a string, how many nodes are there in the mode corresponding to $5f$?

2.15 Do the following series of frequencies correspond to a harmonic series, an overtone series, or both?

(a) 500 Hz, 1020 Hz, 1600 Hz, 2000 Hz, ...

(b) 400 Hz, 800 Hz, 1200 Hz, 1600 Hz, ...

(c) 400 Hz, 1200 Hz, 2000 Hz, 2800 Hz, ...

2.16 A string is vibrating in the mode shown here:

(a) If the length of the string is 37 cm, what is the wavelength of this mode of vibration?

(b) If the fundamental mode has a frequency of 250 Hz, what is the frequency of this mode?

2.17 Consider a 1 m length of string. Sketch the first four modes of vibration. Mark the nodes and antinodes, and find the wavelength, in each case.

2.18 If you lightly touch the string of a violin or cello, the string vibrates in such a way as to have a node at the point that you touch the string.

(a) If you touch the string right in the middle, sketch the first three modes that can vibrate.

(b) What kind of series do the frequencies of these modes form with respect to the lowest frequency in part (a)? That is, is it a full harmonic series, a partial harmonic series, or neither?

2.19 Consider a string on a cello or violin. If you press firmly on the string, you effectively change the length of the string without changing the tension.

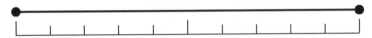

(a) Mark on the string with an X the position where you should press to raise the pitch of the string by a major just third.

(b) Now, if you lightly touch the string, you don't change the length, but you create a node in the vibrational mode at that point. Sketch the mode if you lightly touch the string at the same position as in part (a).

(c) What interval is the note generated in part (b) above the original frequency?

2.20 The strings on a violin, viola, or cello are tuned a fifth apart.

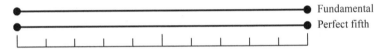

(a) On the figure, indicate with an (x) where you would press your finger on the fundamental string to bring its pitch up a fifth to create a unison between the strings. This technique requires you to place your finger precisely in the correct location.

(b) Alternatively, you can place a finger lightly on each string to produce a harmonic. Indicate with a dot (•) where you would place your finger on each string to produce a common harmonic.

(c) Why is this technique more accurate than the first?

2.21 The vibrational frequency of a string is given by Eq. 2.2:

You can test this formula in the laboratory with the apparatus shown in Fig. 2.13. But, first, one needs to make a prediction from the formula. The string is stretched by hanging a weight from one end of the string. L is the length of the string in meters, U is the mass of 1 m of the string, T is the tension on the string and is given by the mass of the weight, M, in kg times the acceleration of gravity $g = 9.8$ m/s^2. So $T = Mg$. n can be any integer greater than or equal to 1 that gives the different overtones.

(a) If M = 0.4 kg, L = 0.7 m, U = 0.0003 kg/m, find the vibrational frequency of the first four overtones of the string: f_1, f_2, f_3, and f_4.

(b) By how much must we increase or decrease M to double the frequency of the string?

(c) By how much must we increase or decrease M to reduce the frequency by a fifth (3/2)?

Section 2.4

2.22 The Dorian scale results from going up by fifths 3 times and down by fifths 3 times. Going up by fifths, the last note is 27/16. Going down by fifths, the last note we got was 32/27. What are the next two notes that would result by going up and down by fifths once more?

(1) 81/64   (2) 3/2   (3) 32/27   (4) 128/81

(a) 2 and 3
(b) 1 and 2
(c) 3 and 4
(d) 1 and 4

2.23 This is the pentatonic scale, according to the Pythagorean system: 1, 9/8, 4/3, 3/2, 16/9, 2

What are all of the one-step intervals in this scale?

(1) 3/2   (2) 9/8   (3) 32/27   (4) 256/243

(a) 2 and 4
(b) 2 and 3
(c) 1, 2, and 3
(d) 2, 3 and 4

2.24 This is the Aeolian scale, our minor scale, according to the Pythagorean system:

1, 9/8, 32/27, 4/3, 3/2, 128/81, 16/9, 2

What are all of the one-step intervals in this scale?

(1) 3/2   (2) 9/8   (3) 32/27   (4) 256/243

(a) 2 and 4
(b) 2 and 3

(c) 1, 2, and 3

(d) 2, 3, and 4

2.25 Which of these Pythagorean scales do you get when you only go up by perfect fifths and come down by octaves to stay in the original octave?

(a) 1/1, 9/8, 32/27, 4/3, 3/2, 128/81, 16/9

(b) 1/1, 9/8, 81/64, 4/3, 3/2, 27/16, 243/128

(c) 1/1, 9/8, 81/64, 729/512, 3/2, 27/16, 243/128

(d) 1/1, 256/243, 32/27, 4/3, 3/2, 128/81, 16/9

2.26 In the Dorian scale starting on D, what is the interval needed to go from the A (3/2) to the C (16/9)? Express your answer as a ratio, for example, 9/8.

2.27 Starting from $f = 1$, find the Pythagorean scale formed by going down by fifths (3/2) and up octaves (2) to stay between 1 and 2.

2.28 Starting from $f = 1$, find the Pythagorean scale formed by going up by perfect fourths.

Section 2.5

2.29 I have a 30 cm length of string that vibrates at 200 Hz in its fundamental. What is the wavelength of this mode?

(a) 60 cm

(b) 200 cm

(c) 0.005 s

(d) 30 cm

2.30 I have a 30 cm length of string that vibrates at 200 Hz in its fundamental and I place a bridge 1/3 of the distance from one end so that only the long portion can vibrate. What is the new frequency and wavelength of the fundamental? Note, a bridge is like a fret—it simply keeps the short part of the string from vibrating.

(a) 267 Hz, 10 cm

(b) 133 Hz, 40 cm

(c) 300 Hz, 40 cm

(d) 200 Hz, 30 cm

2.31 I have a 30 cm length of string that vibrates at 200 Hz in its fundamental as in Problem 2.30. This time, I only use the bridge when I pluck the string—still 1/3 of the distance from the end. What is the frequency and wavelength now? After I pluck the string, the whole string can vibrate.

(a) 200 Hz, 30 cm

(b) 600 Hz, 20 cm

(c) 133 Hz, 40 cm

(d) 600 Hz, 40 cm

2.32 A wave on a string has a velocity of 175 m/s and vibrates at 420 Hz. What is the length of the string?

(a) 0.208 m

(b) 0.417 m

(c) 2.4 m

(d) Cannot determine the length without knowing what mode the string is vibrating in.

2.33 A 2.8 m length of string in mode number 4 vibrates with a frequency of 441 Hz. What is the velocity of the wave on the string?

2.34 If a tension of 6.3 kg·m/s^2 is applied to a string with a density of 0.0005 kg/m, what is the velocity of the wave on the string?

2.35 A string has a wave velocity of 81 m/s and vibrates at a frequency of 454 Hz. What is the wavelength of the vibration of the string?

2.36 The velocity of a wave on a string is 495 m/s. If the string vibrates in a mode with a wavelength of 0.47 m, first find the frequency of the vibration. From this frequency, find the wavelength of the sound wave produced. Remember, always use 343 m/s for the velocity of sound in air, unless otherwise specified.

2.37 Plot the first three modes of a vibrating string, fixed at both ends. If the length of the string is 2 m, and the velocity of the wave on the string is 200 m/s, what is the wavelength and frequency of each mode?

Section 2.6

2.38 I have a 50 cm-long air column, closed at both ends. It is oscillating in its second mode of vibration. What is the wavelength and frequency of this mode? $v = 343$ m/s.

(a) 343 Hz, 100 cm
(b) 686 Hz, 100 cm
(c) 343 Hz, 200 cm
(d) 686 Hz, 50 cm

2.39 I have a 50 cm long air column, open at both ends. It is oscillating in its third mode of vibration. What is the wavelength and frequency of this mode? $v = 343$ m/s. Don't include the end correction.

(a) 1029 Hz, 33 cm
(b) 686 Hz, 17 cm
(c) 1029 Hz, 150 cm
(d) 343 Hz, 300 cm

2.40 I have a 50 cm long air column, open at one end and closed at the other. It is oscillating in its second mode of vibration. What is the wavelength and frequency of this mode? $v = 343$ m/s. Don't include the end correction.

(a) 229 Hz, 150 cm
(b) 686 Hz, 100 cm
(c) 1029 Hz, 33 cm
(d) 514 Hz, 67 cm

2.41 Which of the following is a possible mode for an air column open at one end and closed at the other? For the correct one, mark the nodes and antinodes and find the wavelength, if the length is 1 m.

2.42 Plot the first three modes of a 26 cm tube with two closed ends. Find the wavelength and corresponding frequencies. Use $v = 343$ m/s.

2.43 Plot the first three modes of a 26 cm tube with one closed end and one open end. Find the wavelength and corresponding frequencies. Use $v = 343$ m/s.

2.44 Sketch the *second* mode of vibration for the following tubes and find the wavelength and frequency of the mode if the tube is 0.5 m long:

(a) An open-open tube.

(b) An open-closed tube.

(c) What is the interval between these two frequencies?

2.45 Plot the first three modes of a 1 m tube with two open ends. Find the wavelength and corresponding frequencies. The velocity of sound is 343 m/s. Do not include the end correction.

2.46 (a) Sketch the first two modes of an open-open air column with a length of 0.35 m. Indicate the nodes (•) and antinodes (x) and find the wavelength and frequency.

(b) Sketch the first two modes of an open-closed air column with a length of 0.35 m. Indicate the nodes (•) and antinodes (x) and find the wavelength and frequency.

2.47 A finger hole in a woodwind instrument creates a node in the mode of vibration at that point. This is just like gently touching a string when plucked to produce an overtone. A finger hole is 10 cm from one end of a 30 cm tube. Which of the following cases will produce an overtone? In other words, which of these modes has a node at the specified point? Indicate all that work. (Hint, there is more than one.)

(a) Closed-closed mode 1

(b) Closed-closed mode 3

(c) Closed-closed mode 4

(d) Open-closed mode 2

(e) Open-closed mode 3

(f) Open-closed mode 4

(g) Open-open mode 1

    (h) Open-open mode 2

    (i) Open-open mode 3

2.48 I have a 50 cm-long air column, closed at both ends. It is oscillating in its second mode of vibration. How many nodes are there?

2.49 We can analyze a tube and determine its resonant frequencies. However, what is really useful is the reverse. If we measure the frequencies we can determine the length of the tube. This will be true even if we cannot see the tube directly (either it is too small or too far away). What length (in meters) of a closed-closed tube will produce the following set of frequencies: 207 Hz, 414 Hz, 621 Hz? Assume the speed of sound is 343 m/s.

2.50 We can analyze a tube and determine its resonant frequencies. However, what is really useful is the reverse. If we measure the frequencies we can determine the length of the tube. This will be true even if we cannot see the tube directly (either it is too small or too far away). What length (in meters) of a closed-closed tube will produce the following set of frequencies: 382 Hz, 573 Hz, 764 Hz? However, I did not record the fundamental, so you will need to determine that, first. Assume the speed of sound is 343 m/s.

2.51 The speed of sound is a function of temperature, given by Eq. 2.8. At 31 °C a flutist plays an "A", at 478 Hz. If the temperature goes up by 9 °C and everything else remains the same (in other words, the wavelength of the note), what is the new pitch?

2.52 I have an open-closed tube with a length of 0.6 m and a radius of 1.7 cm. What is the wavelength of mode number 2, in meters? The speed of sound is 343 m/s.

2.53 I have an open-open tube with a length of 0.7 m and a radius of 1.4 cm. What is the frequency of mode number 3, in Hz? The speed of sound is 343 m/s.

2.54 Note: This problem is probably the most difficult problem of all. It requires you to use all of your knowledge of air columns in one simply stated question.

An air column has a length of 0.50 m. The velocity of sound is 343 m/s. You find that the air column has a resonance at 503.54 Hz. From this you can determine if the air column is closed-closed, open-closed, or open-open. However, the frequency may be slightly off from what you expect because of the end correction at the open end(s). Knowing that the end correction is 0.6 × (radius of the tube) and knowing that a closed-closed tube has no end correction, an open-closed has one end correction, and an open-open has two end corrections, what is the radius of the tube in centimeters?

2.55 You have a tube that is 0.63 m long. A resonance of the tube was measured for each of three different cases: both ends closed, one end open and one end closed, and both ends open. Also, the frequency is affected by the end correction. For each frequency, state whether the tube is open-open, open-closed or closed-closed and what mode it is in. You don't have to calculate the end correction, itself.

(a) 402.0 Hz

(b) 1088.9 Hz

(c) 527.7 Hz

2.56 You start with a closed-closed tube with a fundamental frequency of 571.7 Hz. Assume that the speed of sound is 343 m/s.

(a) How long is the tube?

(b) Now, you unblock the ends, so that you have an open-open tube with the same length as before. What should the radius of the tube be so that the open-open tube has a frequency one half step lower than the closed-closed tube? (use 16/15 for the half step).

2.57 When you measure the resonances of an air column, you find the following series of frequencies: 310, 620, 930, 1240, ... Hz. The tube has a radius of 2 cm.

(a) Is the tube open-open or open-closed?

(b) Find the actual length of the tube, taking into account the end correction. Remember that an open-open tube has an end correction at each end.

2.58 Suppose you measure the following overtones of an air column: 375 Hz, 625 Hz, 875 Hz, 1175 Hz. However, you also realize that you probably did not record the fundamental frequency. Is the air column open-closed, open-open, or is the series wrong? (that is, does not correspond to either case.) Give your reasoning. If it is a proper series, what is the fundamental frequency?

2.59 A finger hole in a woodwind instrument creates a node in the mode of vibration at that point. This is just like gently touching a string when plucked to produce an overtone. A finger hole is 10 cm from one end of a 30 cm tube. In which kind(s) of tube will this produce an overtone: closed-closed, open-closed, open-open. This means that you can find a proper mode with a node at the specified point. Sketch the case(s) that work.

2.60 Which overtones of a closed-closed air column have nodes in the middle of the tube?

Which overtones of an open-open air column have nodes in the middle of the tube?

2.61 A wooden whistle is a simple instrument, as shown in the diagram:

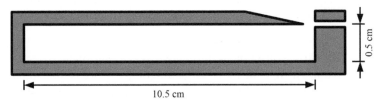

(a) One question that arises is whether the whistle acts as a closed-closed or open-closed air column. There is an opening, but it is off to the side, so the answer is not obvious. However, I measured the pitch of such a whistle to be 816.7 Hz. What would the fundamental frequency of the whistle be for each of the two cases?

(b) Is the whistle, closed-closed or open-closed? Is there an end correction?

(c) Normally, the end correction means that the tube seems to be longer by an amount 0.6 × (radius of tube). If the whistle was open-closed, what would the fundamental frequency be with and without the end correction?

2.62 With real instruments, it is sometimes hard to tell if the instrument acts like an open-open or an open-closed air column. A flute, for example, clearly has one open end, while the other end is closed. However, the blow hole may act like an open end. So, we would like another way to figure this out.

(a) The lowest note on a flute is a "C" at 262 Hz. The length of the flute is about 24 inches. From this information, determine whether the flute acts like an open-open or open-closed air column. Does your answer predict the correct overtone series for a flute?

(b) The length of a clarinet is also about 24 inches, but the lowest note on a clarinet is 147 Hz. Does the clarinet act like an open-open or open-closed air column and why? Does this agree with the overtone series?

2.63 The following series is missing one or more of the lowest frequencies:

$$2500, 3500, 4500, 5500 \text{ Hz}, \dots$$

(a) This series can be part of the overtone series for an air column with which of the following boundary conditions (there may be more than one): closed-closed ends, closed-open ends, open-open ends?

(b) What are the missing low frequencies?

2.64 The following series is missing one or more of the lowest frequencies:

$$5000, 7000, 9000, 11000 \text{ Hz}, \dots$$

(a) With which of the following boundary conditions (there may be more than one) can this series be part of the overtone series for an air column: closed-closed ends, closed-open ends, open-open ends?

(b) What are the missing low frequencies?

2.65 Consider building a pan flute. The radius of the tubes is 1 cm. One end of each tube is closed with a rubber stopper. The question is, do you need to worry about the end correction?

2 cm

(a) First, consider just a regular open-closed tube with no end correction. How long should we make the tube so that it plays an A = 440 Hz?

(b) How far into the tube must the stopper be pushed to raise the pitch by exactly a just whole step (9/8) compared to part (a)? Is you answer reasonable?

(c) Going back to part (a), now find what the pitch would be if we did consider the end correction.

2.66 State whether the following series correspond to an open-open tube, an open-closed tube, or neither. One or more of the low frequencies may be missing. If a series does correspond to an open-open or an open-closed tube, give the fundamental frequency of the tube.

	Type of tube	Fundamental
300, 400, 500, 600 Hz ...		
600, 1000, 1400, 1800 Hz ...		
100, 200, 400, 800 Hz ...		
120, 220, 320, 420 Hz ...		

2.67 Each of these overtone series corresponds to an open-open or open-closed air column. However, some low frequencies may not have been recorded. Also, there is exactly one extra or one missing frequency. In each case, give that extra frequency, determine whether the tube is open-open or open-closed, and give the fundamental.

	Extra frequency	Type of tube	Fundamental
375, 625, 700, 875, 1125 Hz ...			
300, 450, 750, 900, 1050 Hz ...			
450, 675, 900, 1125, 1300 Hz ...			
800, 960, 1600, 2240, 2880 Hz ...			

Section 2.7

2.68 The Pythagorean scales are not used today in Western music, because:

(a) There are other ways to create the musical scale.

(b) The planets do not actually follow the "Music of the Spheres."

(c) The Pythagorean system was not properly understood in the Middle Ages.

(d) You cannot build an instrument according to the Pythagorean system.

2.69 The Medieval woodcut showing the ideas of Pythagoras is significant because:

(a) It shows that the physics of bells and water glasses was understood by the Middle Ages.

(b) It shows that in the Middle Ages, the ideas of the ancient Greeks were simply accepted and applied to other situations without testing.

(c) It explains why the planets follow the Pythagorean musical scale.

(d) All of the above.

2.70 Look at the lower left-hand panel in the woodcut describing the Pythagorean ratios.

(a) First assume that the numbers give the relative LENGTHS of the strings. Give the interval of the strings (6) through (16) with respect to string (4).

(b) Now interpret the drawing literally—that is, the numbers give the relative WEIGHTS on the strings. Again, give the interval that the strings (6) through (16) in that order make with respect to string (4). If the frequency is exactly equal to an interval that we know, give the name of the interval. Otherwise, give the ratio with three decimal points (for example, 1.234).

Section 2.8

2.71 Discuss one strength and one weakness of each of these tuning systems: Pythagorean, just, Equal Temper.

2.72 We have discussed intervals of an octave, a fifth, and a fourth. Of course, there are more intervals in music. In particular, the major third. The major third corresponds to going up two whole steps.

(a) In the Pythagorean scale, what is the ratio of frequencies corresponding to a major third?

(b) The answer to part (a) is not a very simple fraction and some philosophers did not like this. What simple ratio of integers less than 10 is the closest to the Pythagorean major third? (This is called the just scale.)

(c) On a piano, a half step corresponds to a frequency ratio of 1.0595 and two half steps exactly equals a whole step. On a piano, what would be a major third?

2.73 (a) The minor third on the piano corresponds to three half steps. What is the equal temper minor third?

(b) Pythagoras would create a minor third by going up a fifth and coming down by a Pythagorean third. What is the ratio for the Pythagorean minor third? (Give the answer as a ratio of two numbers.)

(c) Pythagoras could also have formed the minor third by going up a fourth twice and coming down by a fifth. Does this give the same ratio as part (a)? (Show your work!)

2.74 (a) Why does 1.0595 correspond to a half step on a piano?

(b) Pythagoras would create a major sixth by going up a fifth twice and coming down by a fourth. What is this ratio?

(c) What would a major sixth (nine half steps) correspond to on a piano?

(d) The just interval would be the closest ratio of small integers to the piano interval. What would a major sixth be in the Just scale?

2.75 (a) The minor sixth on the piano corresponds to eight half steps. What is the equal temper minor sixth?

(b) Pythagoras would create a minor sixth by going DOWN by a perfect fifth FOUR times and coming up by octaves. What is the ratio for the Pythagorean minor sixth? (This ratio should be very close to the answer from part (a).

(c) The just minor sixth is the ratio of two numbers, each less than 10, that is the closest to the answer in part (a). What is the just minor sixth?

2.76 The overtone series of a brass instrument is a harmonic series. If the fundamental frequency is f, the overtone series is 1*f*, 2*f*, 3*f*, 4*f*, 5*f*, 6*f*, 7*f*, 8*f*, ... What combination of intervals from the fundamental does each of these tones correspond to? For example, 4*f* is two octaves above the fundamental.

2.77 We considered the intervals formed by the first seven overtones of a brass instrument, 1*f*, 2*f*, 3*f*, 4*f*, 5*f*, 6*f*, 7*f*. Now consider the next few: 8*f*, 9*f*, 10*f*, 11*f*, 12*f*. What interval does each overtone form with respect to the 8*f* overtone? If the interval is not exactly one of the intervals you know, which one is it closest to?

2.78 There are several intervals that we have not considered, yet. For each interval, find the ratio corresponding to that interval in the Equal Temper scale. Then find the ratio of integers, which is close to this value—this will determine the interval in the just scale:

Minor third = 3 half steps
Minor sixth = 8 half steps
Major sixth = 9 half steps
Major Seventh = 11 half steps

2.79 Show that certain pairs of intervals combine to give an octave. How many pairs can you find? Some pairs may not come out perfectly in the just scale or the Pythagorean scale. Give the pairs that come close and state whether they give a perfect octave or not in each of the scales. Why will the pairs always come out perfectly in the Equal Temper scale?

2.80 Triads consist of three notes. They contain three different intervals: the first to second note, the first to third, and the second to third.

(a) The major triad has notes in the ratio 4:5:6. What are the three intervals in this triad? This is the triad that Ptolemy liked.

(b) Notes in the ratio 3:4:5 also make up a triad. What are the three intervals in this triad?

(c) (Not required, not for credit) What musical chord does (b) correspond to?

2.81 The Pythagorean pentatonic scale consists of the following intervals:

1   9/8   4/3   3/2   16/9   2

(a) Scales are generally formed from large steps and small steps that connect the adjacent notes in the scale. For example, the step from 1 to 9/8 is simply the interval 9/8. Find the next four steps of the pentatonic scale.

(b) In part (a), you should end up with only two intervals, a small and a large step. Find the nearest just intervals to each of these steps. Remember, the just intervals are ratios of integers less than 10 (in most cases). Note, one of both steps might already be a just interval.

Chapter 3

Non-Melodic Instruments: Non-Harmonic and 2D Systems

3.1 Vibrating Bars and Non-linear Dependencies

In Chapter 2, we only considered systems with harmonic overtones: the vibrating string and the air column. However, these are rather special, although important, cases. Many systems, including some musical instruments, do not have harmonic overtones.

Unlike strings, which vibrate because they are under tension, some instruments are based solely on stiffness, like the vibrating bars of a xylophone or a marimba. In this case, no tension needs to be applied to a bar for it to vibrate. For these instruments, the vibrational frequency depends on the stiffness of the bar, not the tension applied.

Yet another class of instruments depends on vibrations in two dimensions. This class includes many percussion instruments, like drums or cymbals. Here, the vibrating part of the instrument is a surface. We will return to this situation in a later section.

A common feature of both vibrating bars and two-dimensional instruments is that they are not so melodic and are generally used as percussion instruments. Why this is so will be discussed in the next section. Finally, note that some instruments,

The Music of Physics: An Introduction to the Harmonies of Nature
George N. Gibson
Copyright © 2024 Jenny Stanford Publishing Pte. Ltd.
ISBN 978-981-4968-99-7 (Hardcover), 978-1-003-48481-3 (eBook)
www.jennystanford.com

like cymbals, for example, vibrate due their stiffness and are two dimensional. As a result, these systems are much harder to analyze quantitatively.

Modes of Vibration of a Bar

What does a bar look like when it vibrates? The answer to this depends partially on how the bar is held or supported. To start with, let's just assume that we are on the space shuttle and the bar is floating freely in space:

If it starts vibrating, it will look something like this:

The vibrational modes of a bar can get quite complex. It can bend in the long direction, bend in the short direction or it can twist. Vibrating bars for instruments are made long and narrow so that the main bending is in the long direction. Twisting and the short direction are not so important because they will lead to vibrational frequencies which are much higher in pitch. So, we will focus on just one long direction. In this case, it is easier to look at the bar from the side. Figure 3.1 shows the first few modes of vibration.

The exact shape of these modes cannot be determined without calculus, which is beyond the scope of this course, although there are some features that we can recognize. There are still nodes and antinodes and there is one more node for each higher mode of vibration. However, the spacing between the nodes and antinodes is not the same. To get some sense of what the shape should be, just remember that the bar is very stiff and does not want to bend, so any part that is not forced to bend will be straight.

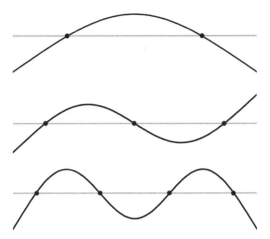

Figure 3.1 Modes of a free-free vibrating bar with the nodes marked.

Notice that the very ends of the bars always end up straight. Calculating the position of the nodes again requires calculus, but for the fundamental, the nodes are 22% of the length of the bar from each end (this can be verified with Lab 3.1).

At this point, we usually write down the frequency of each mode relative to the fundamental. Again, there is no simple way to figure out the frequency of the higher modes without calculus, so we look them up in a book and find that the overtones series is:

$$1f, 2.76f, 5.40f, 8.93f, \dots \tag{3.1}$$

This is certainly not a simple harmonic series, but it can easily be measured in a lab (also part of Lab 3.1).

Just as with an air column, we can have different boundary conditions at the ends of the bar: the ends can be free to vibrate, or they can be clamped. If the end is clamped, there will be a node, and if it is free, there will be an antinode. Furthermore, these can be combined in various ways, just like the air column: free-free, free-clamped, and clamped-clamped. This last one, clamped-clamped is highly damped and not that useful. However, the free-clamped combination has an interesting application. Figure 3.2 shows what its modes look like.

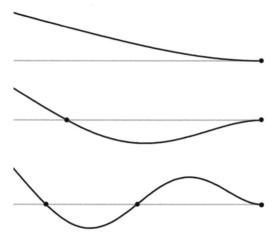

Figure 3.2 Modes of a free-clamped vibrating bar.

The overtones series for this arrangement is: 1*f*, 6.27*f*, 17.6*f*, ... Notice how much higher the first overtone is compared to the fundamental. Here, the shape of the modes is clearly much different from a vibrating string or air column. Actually, if you have ever tried to twirl a chain or heavy string around, you may notice that it takes on the same modes as the free-clamped bar.

Dependence of Frequency on Length

Once we have the fundamental of a vibrating bar, we can predict the frequency of the higher modes using Eq. 3.1, but what does the fundamental depend on? It depends on the thickness of the bar, the material, the shape, and the length. The actual frequency of the fundamental of a vibrating bar is much harder to write down than for a vibrating string, so we will not do this. Rather, we will focus on just one of the dependences: how the fundamental frequency changes with the length of the bar.

Remember that for a string, if you double the length, the frequency goes down by a factor of 2. So, the frequency, *f*, depends on length, *L*, in the following way: f_{string} = constant/*L*. We would say that the frequency in linearly but inversely proportional to the length. Also, remember that the string had harmonic overtones and the distance between nodes is always the same, in a particular mode.

As we have seen, the overtones of a bar are not harmonic and the modes are more complicated than for a string. So, it is reasonable to ask whether the fundamental of the bar will vary with length in the same way that the string does. In fact, the fundamental does not have the same dependence on length. We only need to look at a xylophone (Fig. 3.3).

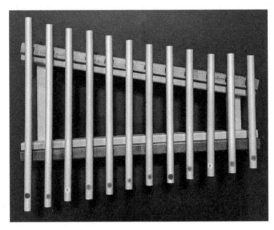

Figure 3.3 Xylophone showing how the lengths are not proportional to the frequency of the notes.

If we look at a single octave of the xylophone, the frequency changes by a factor of 2. But if we look at the lengths of the bars an octave apart, the upper note is not close to one half of the length of the lower note. It is longer than we would expect, close to three quarters of the length of the lower note. If we carefully measure the lengths of the bars, we would find that they are in a ratio of 0.707 to 1. So, the shorter the bar the higher the frequency, just like a string, but where does the 0.707 come from? As it turns out, $(0.707)^2 = 0.5 = 1/2$, which is the ratio of the frequencies. So, it turns out that the frequency depends on the *square* of the length. If you change the length by a factor of 0.707, the frequency changes by a factor of $(0.707)^2$. How can we express this as a relationship? We write:

$$f_{\text{bar}} = \frac{\text{constant}}{L^2} \tag{3.2}$$

Here, the constant hides all the other complexities of determining the frequency of the bar. This relationship simply highlights the dependence on length. Again, this dependence of frequency on length is readily measured (Lab 3.1).

Real xylophones and marimbas are yet more complicated. It turns out that instrument makers shape the bars by thinning them in the middle, to change the overtone series. This can make the bars sound a little better, but this also changes the pitch of the fundamental. So, the dependence on length does not hold perfectly.

Worked Example

Consider a metal bar 20 cm long and you want to make a second bar a perfect fifth higher in pitch? How long should you make the bar?

Solution

Since, f = constant/L^2 for a vibrating bar, the qualitative dependence says that the bar needs to be shorter, since the length is in the denominator. Also, the perfect fifth is a factor of 3/2 higher in pitch. If you treated the bar like a string, you would shorten the bar by the same factor of 3/2. However, the bar depends on L^2, so changing the length by 3/2 would change the frequency by $(3/2)^2$ = 9/4. So, you need to change the length by a factor that, when *squared*, gives 3/2. This will be $\sqrt{3/2}$ = 1.225. Thus, the new bar should have a length equal to (20 cm)/1.225 = 16.33 cm.

3.2 Overtone Series, Addition of Waves, and Tone Quality

We have, so far, left out a very significant point. We have discussed the fundamental frequency of a vibrating system and its scaling with length. We also are aware of the fact that there are many possible modes of vibration each with its own vibrational frequency. This set of frequencies is called an overtone series and can be a harmonic, a partial harmonic, or a non-harmonic, series.

The key point, here, is when a note is sounded on an instrument, *all* of the different modes of the instrument can start vibrating, and each mode contributes its own frequency to the overall sound produced. This is what makes each instrument sound different, or what gives each instrument its own tone quality or timbre.

However, it gets even a bit more complicated, because each mode does not necessarily contribute the same amount of sound, even for the same note on the same instrument. If you pluck a guitar right in the middle of the string, it will sound quite different than if you pluck the string very close to one end of the string. Even though it is the same instrument playing the same note, the two ways of playing the note can sound different. A violin can be plucked or bowed. These two ways of playing the note, again, sound very different, even if the pitch is the same and the instrument is the same. What is changing is the amount of each overtone contributing to the tone quality. Lab 1.3 shows how to produce different timbres.

Addition of Waves

Back in Section 1.1, we discussed the fact that the shape of a wave determines its tone quality, yet, now we are saying that it is the overtones that determine the tone quality. What is going on? Well, these are actually the same statement and reveal a fundamental connection.

A pure note, one with no overtones, has a shape that is a sine wave (Fig. 3.4A). A pure note with twice the frequency, the second harmonic, is also a sine wave (Fig. 3.4B). Now, let's consider that we are playing a note that contains the second harmonic, in addition to the fundamental. What does the shape look like? All that we have to do is add the two waves together as shown in Fig. 3.4C.

There are several important aspects to this new shape. First, its shape is not a simple sine wave, so it will sound richer than just a pure tone. Second, *even though the shape is different, the period and frequency are the same as the fundamental.* You can see this by counting the number of full oscillations in each. The reason is that the second harmonic has twice the frequency, but it also repeats with the period of the fundamental. This is very important, because it means that the pitch of the combined wave stays the same. So, now we can see the connection between shape, harmonics, and tone quality. If we add harmonics to the fundamental, we change the shape of the wave, but not its pitch, so this gives us a way of independently controlling the tone quality without affecting the pitch. It does not matter if we add a wave at 3*f*, 7*f*, or 32*f*. The shape will change but the pitch will not. This relates back to Lab 1.3 from Chapter 1. But now, we more fully understand the connection between shape, harmonics, and tone quality.

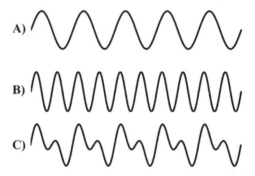

Figure 3.4 (A) Fundamental; (B) second harmonic; (C) sum of (A) and (B).

Melodic vs. Non-Melodic Instruments

Above, we only considered the effect of adding a harmonic overtone to the fundamental, because there is a crucial difference if the overtone is harmonic or not. If the overtone is harmonic, the pitch does not change. Let's see what happens if we add an overtone which is not harmonic, for example, the first overtone of a vibrating bar at 2.76*f* to the fundamental (Fig. 3.5).

As we can see, the summed wave (C) does not repeat with a well-defined period. If you listen to this sound, it will not have a clear pitch. You will hear both the fundamental and the overtone, but they do not support each other in establishing a pitch (Lab 3.2). A sound without a distinct pitch is usually regarded as unpleasant.

So, an instrument with harmonic overtones means that you can add the harmonics in various amounts to adjust the tone quality while preserving a sense of pitch and producing a pleasing sound. A skilled player can vary the mixture of harmonics to change the tone quality to suit the mood of the music, whether it is bright and clear, mellow, or dark and rich.

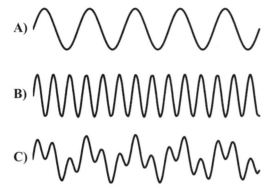

Figure 3.5 (A) Fundamental; (B) first overtone of a vibrating bar; (C) sum of (A) and (B).

This is not possible with a non-harmonic instrument. For the most part, the overtones must be suppressed, or an unpleasant sound without a strong pitch will be produced. Of course, sometimes, this is what is called for. Cymbals, for example, fall into this category. They are used for sound effects, but not for pitch, melody, or harmony.

An instrument like a xylophone can be used to play a melody, but this only works because the overtones are greatly suppressed—for the most part, you only hear the fundamental. In addition, bars can be shaped, as discussed above, to bring the overtones closer to harmonics. While this avoids the harshness of non-harmonic overtones, there is almost no ability to change the tone quality. Thus, the range of expression is greatly restricted in this type of instrument. This is why they are still relegated to the percussion section.

Drums are like cymbals, in that they are mostly used for effect, not melody or harmony. One exception is the tympani (or kettledrum). Here great effort goes into designing tympani to bring their overtones close to harmonics. When this is achieved, the drum gains a sense of pitch. This is also why symphony tympani are so expensive.

One way the overtone mixture can be controlled in a vibrating bar is by how the bar is supported. To hear the fundamental, the ends must be free to vibrate. As a consequence, the bar cannot be held at the ends. However, since the nodes are positions where the bar does not vibrate, it can be held at those points without damping the vibration. This point is about 22% of the length of the bar from each end and you can see this on any xylophone by looking at how the bar is supported. However, these points are not nodes for the higher modes of vibration, so all of the higher modes will be damped to varying degrees. Thus, how the bars are held gives another method for bringing out the fundamental while suppressing the higher modes. What would happen if a bar was held at just one point exactly in the middle of the bar?

Overtones and Instruments

The overtones also affect the construction of instruments. The last section gave an example of this where it was mentioned that the shape of a vibrating bar could affect the overtone series. The first overtone of a bar is at $2.76f$. This is not close to any harmonic or even any of the intervals that we know. However, if we go up an octave and a perfect fourth from the fundamental we get to $(2)(4/3)f = (8/3)f = 2.67f$. So, if we could adjust the first overtone of the bar slightly and bring it down from 2.76 to 2.67, the bar would sound much better. Thinning the bar somewhat in the middle of the bar achieves this adjustment. Along the same lines, we will discuss two other examples: grand pianos and tuning forks.

Strings on a Grand Piano

When we discussed vibrating strings we assumed that the string is perfectly flexible. In other words, it has no stiffness. A piece of thread or dental floss falls in this category. The strings on a guitar are also fairly flexible while the violin strings have a bit of stiffness. The most extreme case is a piano string. Piano strings are so stiff they are rather hard to work with. As we will see, this changes the overtone series and affects the tone quality of a piano in addition to how it is tuned.

A very thin metal wire is very flexible and behaves like an idea string. For example, the first overtone is at the harmonic frequency $2f$. Now, as the wire gets thicker, it takes on some stiffness. This has the effect of making the overtones higher than they would be for a perfectly flexible string. The musical term for this is that the overtones become sharper. The stiffer the string or wire, the sharper the overtones. In the limit that the wire becomes so thick that you would consider it a bar instead of a string or wire, the first overtone would be at $2.76f$ not $2f$.

On a piano, because of the thick wire strings, the overtones are all slightly sharp. It is noticeable and a good piano tuner will correct for this. Say you first tune middle C to its standard pitch, which we will call f. The C three octaves higher should be at $(2)^3 = 8f$. The 7th overtone (or 8th harmonic) of middle C should also be $8f$. However, because of the stiffness, it is more like $8.05f$ (as measured on an actual grand piano, see Lab 3.3). Because many of the overtones sound when you strike a note on a piano, it is more important to tune the third octave above middle C to the 7th overtone of middle C, rather than making it exactly three octaves higher. This way of tuning is called "stretching the octaves" because each octave becomes slightly larger than it would normally be. This happens naturally if a piano is tuned by ear, because the octaves sound better if the overtones are in tune.

Of course, there is a limit to this. If the overtones are too sharp, the octaves will deviate too much from their ideal value. So, this, in turn, limits the size of a piano. Full concert grand pianos are the largest instruments. The goal is to use as long a string as possible, because this will make the piano louder and the tone richer. However, if you keep making the strings longer, either the tension must be increased or the strings must be made thinner to keep the pitch of the string the same, according to the relationship in Eq. 1.2.

The problem is that thinner strings will break. Therefore, the tension must be increased, which will also break the strings. So, in the end, the strings must be made thicker. Although the thicker string will now decrease the frequency, this is more than compensated by the increase in tension made possible by the thicker string. Ultimately, though, the strings become so stiff that the overtones will be too far out of tune to sound good. In fact, modern grand pianos have reached their limit in size, unless, of course, new material, which is stronger, but not stiffer, can be found.

Tuning Forks

A string has many harmonic overtones giving a warm and rich tone quality. However, having all of the harmonics present does make the pitch a little bit less distinct. So, it is harder to tune an instrument precisely to a violin, or any other instrument. Tuning forks, a metal bar in the shape of a "U", offers a useful alternative. The two lowest modes of vibration are shown in Fig. 3.6.

This clever arrangement makes each "tine" of the tuning fork act like a metal bar clamped at one end. We examined the possible modes in this case, in the previous section. In the figure, we see that the next higher mode is at 6.27*f*. This is so high that the ear can ignore it and focus on just the fundamental. This provides a pure and precise pitch to which one can tune an instrument. Moreover, all tuning forks will sound the same, as there is no difference in overtone pattern from one tuning fork to another.

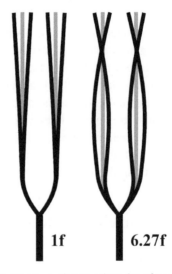

1f   6.27f

Figure 3.6 First two vibrational modes of a tuning fork.

3.3 Dimensional Acoustic Cavities

For the most part, we have been considering waves in one dimension—along a string, down a tube, along a metal bar. Of course, in reality, everything moves in three dimensions. However, it is a bit easier to first consider what changes as we go from one to two dimensions. Also, as mentioned before, there are some musical instruments based on 2D vibrations, including the cymbals and drums. In this section, we will focus on sound waves in two dimensions, as sound waves are the simplest waves to extend to two dimensions.

When we discussed sound waves in 1D, we considered a tube. In this case, the diameter of the tube was much smaller than the length of the tube, so it was pretty clear that the sound waves would travel up and down the length of the tube:

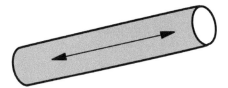

To define a 2D wave, we need a thin rectangular shape:

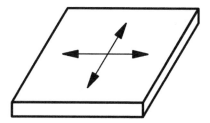

This is called an acoustic cavity. A cavity is anything that confines waves to a well-defined geometry. Here, the waves again travel in the long directions, and, thus, are restricted to a 2D plane. From now on, it will be easier to look at the cavity from the top (Fig. 3.7).

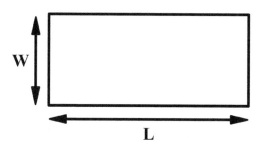

Figure 3.7 Thin rectangular cavity.

When we discussed the air column, we always took L to be the length of the tube. Here, we need to specify two dimensions, the length, L, and the width, W. To start our analysis, we can have the same modes in each direction as we had for a tube. In this case, we will assume that the cavity is closed all the way around corresponding to closed-closed boundary conditions.

Along the length of the cavity, the modes look just like ones Fig. 2.8. The wavelengths will be $2L$, L, $2L/3$, $L/2$, ... and the frequencies are given by Eq. 2.5.

Along the width, we have modes where the wavelengths are $2W$, W, $2W/3$, ... and frequencies are $g_m = mv/(2W)$. We use m here, because we can have different modes in each direction.

So far, the 2D cavity looks just the same as two 1D cavities put together. However, the new feature is that the cavity can have both modes going at the same time. Then the question is, what is the frequency of the overall mode? For example, if we are in the 3rd mode along the length, its frequency is $f_3 = 3v/(2L)$. If we are in the 5th mode along the width, then $g_5 = 5v/(2W)$. What is the frequency of the mode consisting of both of these modes simultaneously? This is called a *combination mode* and it needs two numbers to specify it, n and m, in this example, 3 and 5. The frequency of the combined mode is designated $F_{n,m}$ and is obtained by adding the frequencies of the two separate modes according to the Pythagorean theorem:

$$F_{n,m} = \sqrt{f_n^2 + g_m^2} \qquad (3.3)$$

Now that we can find the frequencies of the modes, we need to know what they look like. Let's start with a simple mode along one direction—the second mode in the long dimension. The mode looks just like a mode along a tube, but spread out in the other direction (Fig. 3.8).

We see antinodes at the edges, since it is a closed boundary condition. Black is the extremes, so that represents antinodes. White shows where the nodes are. Since we have the second mode, there should be two nodes. The upper surface shows the amplitude of the wave, while the lower surface is a two-dimensional representation of the wave.

Figure 3.8 View of the second mode of a thin rectangular cavity in the long direction.

Here is what a simple mode in the other directions looks like—the first mode along the short dimension (Fig. 3.9).

Figure 3.9 Fundamental mode of a thin rectangular cavity in the short direction.

Again, we see antinodes at the edges and one node. So far, this is not too different from an air column—the mode is just stretched out in the other dimension. So, the question is: what happens if we have both modes at once? Mathematically, the two modes get multiplied together, but you will understand better by just looking at the mode (Fig. 3.10).

Figure 3.10 Combination mode consisting of Figs. 3.8 and 3.9.

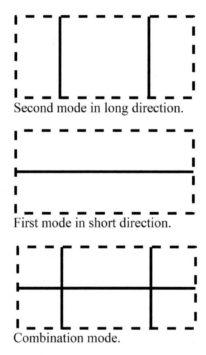

Second mode in long direction.

First mode in short direction.

Combination mode.

Figure 3.11 Alternative representation of cavity modes.

The pattern is now somewhat more complicated, but certain elements are apparent. In 1D, the most important feature in determining the shape of the modes was the location of the nodes.

The same is true in 2D with one difference. In the three images above, white represents the nodes. However, the nodes are no longer points, as they were in 1D, the nodes are now lines and are referred to as *nodal lines*. This is not so surprising—going from one to two dimensions involves adding a dimension. So, if you start with a point and add a dimension, you get a line. The nodal lines are useful in seeing the mode, because we simply keep the nodal lines from each mode. There is a simpler representation of all three modes above, showing just the nodal lines (Fig. 3.11).

Notice that in these drawings, the rectangle is a dotted line. This is because the edges of the cavity are *not* nodal lines. The thick solid lines represent the nodal lines. To completely show the mode, we can also draw the individual modes to the side of the rectangle as shown in Fig. 3.12.

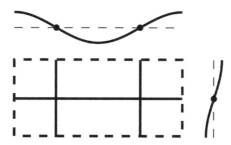

Figure 3.12 Full representation of a cavity mode of a thin rectangle.

Finally, if we know the dimensions of the cavity, we can determine the frequency of the mode. This is a complete description of the combination mode. Such cavities can be studied with Labs 3.4 and 3.5.

As usual, one problem with sound waves is that we cannot see them, so things like nodes and antinodes are a little abstract. There is a way to see nodal lines in 2D waves using what are called "Chladni plates". A Chladni plate is a thin plate of metal cut into any shape, but usually a rectangle or a circle. A mechanical oscillator vibrates the plate at a resonant frequency. Finally, fine sand is sprinkled on the plate. Since the plate is vibrating, the sand dances around and eventually falls off of the plate. However, if the sand is sitting right on a nodal line, the sand does not move, since there is no motion on the nodal line. So, the sand that stays on the plate collects on the nodal lines, showing them quite clearly. Figure 3.13 shows some images of Chladni plates in different modes of vibration.

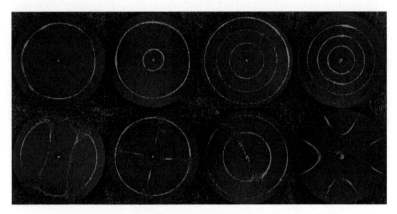

Figure 3.13 Various vibrational modes of a thin metal plate.

Now, don't worry, you will not have to be able to draw these nodal lines. The nodal lines on the Chladni plates are quite a bit more complex than for sound waves in a rectangle. This is just like the fact that the modes of vibrating bars are more complex than for air columns. As mentioned above, vibrating metal plates, like cymbals, are even harder to describe than vibrating metal bars. However, the Chladni plates are still interesting because they vividly show the reality of the nodal lines for 2D waves.

Worked Example

Consider a closed rectangular cavity 30 cm by 20 cm which is oscillating in the third mode in the long direction and the first mode in the short direction. Sketch the individual modes and the nodal lines of the combined mode. Calculate the frequency of the combined mode.

Solution

Since the cavity is closed, there will be antinodes at the boundaries. The third mode will have 3 nodes and the first mode will have 1 node. So, the modes will look like the following:

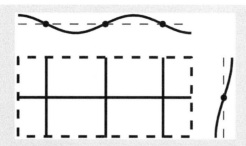

The frequency in the long direction will be:

$$f_3 = 3 \cdot \frac{343\,\text{m/s}}{2 \cdot 0.3\,\text{m}} = 1715 \text{ m.}$$

The frequency in the short direction will be:

$$g_1 = 1 \cdot \frac{343\,\text{m/s}}{2 \cdot 0.2\,\text{m}} = 857.5 \text{ m.}$$

Finally, the combination frequency will be:

$$F_{3,1} = \sqrt{f_3^2 + g_1^2} = 1917 \text{ Hz.}$$

3.4   Cavity Mode Diagrams

As we have seen, waves in two dimensions get significantly more complicated than in one dimension, so now imagine going to three dimensions. The situation gets even more complicated and we need techniques to organize and visualize what is going on. Such methods of organizing data are critical in all areas of physics, as it is important to present information in a way that captures the big picture. As an example of this, we will present the results for the rectangular cavities from the last section in a rather different way. In the last section, we started with a theoretical framework based on our understanding of sound waves in a 1D closed-closed tube and generalized that to 2D. But, what if we knew nothing about tubes? How could we approach understanding the 2D waves in a rectangular cavity? A common approach in physics is to simply gather more data and try to make sense of it. The tricky part of investigating new things is knowing what data to gather and it sometimes takes many attempts.

In this case, let's consider a rectangular cavity in which you can change one of the dimensions easily, using an arrangement shown in Fig. 3.14.

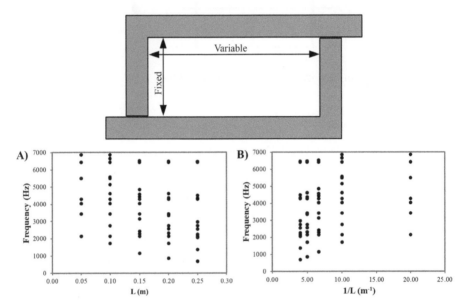

Figure 3.14 Mode frequencies of a rectangular cavity with one variable side measured at various lengths. (A) Frequencies plotted versus the length L. (B) Frequencies plotted versus $1/L$.

If the variable length is called "L", we can plot all of the cavity mode frequencies as a function of L, by measuring them experimentally. In this way, we don't need to know anything about the physics of the rectangle. The results are shown in Fig. 3.14A.

Figure 3.14A has a great deal of information, but it is not so clear what is going on. However, one feature is quite evident: as the length gets shorter, the frequencies tend to get higher. This is actually a fairly general trend for almost anything that vibrates—the smaller the size the higher the frequency. This is an inverse relationship, which we have seen with strings, air columns, and vibrating bars, so it might makes sense to plot the frequencies as a function of one over the length: 1/length (Fig. 3.14B). When we do this, we see that many points can be connected with straight lines.

Now it is possible to start grouping the various lines together into several types (Fig. 3.15). First, there are the lines that do not change with the length (solid grey). These clearly must be modes that are independent of the changing side, and, hence are modes in the fixed direction. Second, there are lines that are straight (solid black) but with a constant slope as a function of $1/L$. This should remind us of the modes of an air column (Eq. 2.5). From these two observations, we can conclude that the rectangular cavity supports the simple modes of an air column in each direction. Finally, there are curved lines (the dashed lines), which indicate something new and more complicated. They depend on the changing length but are grouped along the constant lines indicating that they are somehow a combination of modes in each direction. How they are combined is not immediately obvious from this graph but the important point is that we have learned a lot about rectangular cavities just from this experimental data.

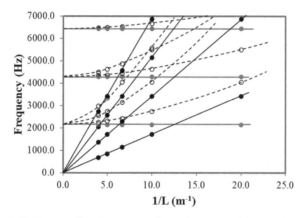

Figure 3.15 Frequencies of a rectangular cavity grouped into various types.

This method of organizing data to understand new physical phenomena is used extensively in research. Figure 3.16 shows data for a very different situation, namely the energy levels of a hydrogen atom in an external electric field. While the details are not important here, the data group into similar patterns:

horizontal lines (grey) that do not change; straight lines with a slope (black); and curved lines (dotted). This kind of graph can help us make sense of what is going on in the experiment.

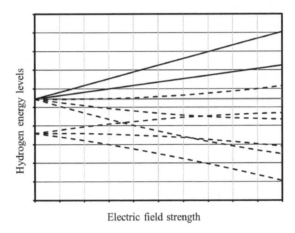

Figure 3.16 Energy levels of a hydrogen atom in an electric field.

3.5 Spectroscopy, Modern Physics, and Music of the Spheres

We have now spent quite a bit of time studying waves and some history of physics and music, so it is time to stop for a moment and ask: what have we learned and why is it important? We will take the rectangular cavities as an example of what this is all about. Lab 3.5 demonstrates how one can study real cavities in a precise way. By identifying the resonances and matching their frequencies with those predicted from a theory, it is possible to obtain a precise measurement of the speed of sound in the cavity. However, it turns out that the speed of sound depends on the temperature in the cavity. So, in the end, we find that we can measure something quite different than what we started out to: the temperature. Thus, the acoustic cavity can actually be used as a thermometer. This is often the case in physics. If you measure something very carefully and try to construct a theory to explain it, you often find that you learn about something that was seemingly unrelated. This is how science progresses from what is known to what is unknown. This is what also leads to many useful and important applications of science. Understanding the frequencies of an acoustic cavity led to a method for measuring the temperature, but, there are many more possibilities.

Perhaps we know the temperature from a normal thermometer, but we don't know what kind of gas is in the cavity. However, some other scientist might have made a catalog of the speed of sound in many kinds of gas. By measuring the speed of sound in the cavity through identifying the resonant frequencies, we could determine the kind of gas in the cavity.

We can also proceed in a different direction. In this case, let's assume we know the speed of sound in the cavity, because we know the kind of gas and its temperature. By measuring the resonant frequencies, we can determine the dimensions of the cavity. Of course, we could have just measured the dimensions with a ruler, but measuring frequencies presents an interesting alternative. Why is this so important? There are many situations where a ruler will not work, but we can still measure frequencies:

(1) We can measure things that are too small to see. For example, we cannot measure the size of atoms and molecules with a ruler. We cannot even see them in a microscope. So how can we measure their size? Atoms and molecules act as little cavities that can be probed with light waves. The resonant frequencies of the atoms and molecules can be measured. To turn these measurements into dimensions of the cavity we need a theory to go along with the measurement. This theory is called quantum mechanics. With the theory and measurement of frequencies, we can then determine the size of atoms and molecules.

(2) We can measure things that are far away. For example, we can learn about astronomical objects, such as stars, galaxies, binary stars, and the expansion of the Universe, by measuring the frequencies emitted by these objects. Indeed, we can weigh distant stars and determine the size of the Universe with similar techniques. An important technique in astronomy is measuring the Doppler shift. This will be discussed in detail in Chapter 5, but basically, the Doppler shift changes the frequency produced by a moving object and is used to measure the velocity of an object. This is how astronomers can determine that the Universe is expanding or how fast galaxies are rotating.

(3) We can measure things that we don't want to get near—an excellent example is spying. During the Cold War, the Soviet government presented a fancy plaque to the American Embassy in Moscow. The Americans were, of course, extremely suspicious of this gift, so they examined the plaque, took x-rays of it, etc., but could not find anything wrong with it. So, they hung the plaque up in the embassy. A few weeks later, the Americans noticed that the Soviets were beaming microwave radiation at the embassy, but they could not figure out why. Finally, a scientist put it all together. The microwaves were probing a cavity in the plaque that had a flexible membrane stretched over it. As people talked in the embassy, the membrane would vibrate, changing (very slightly) the dimensions of the cavity. The microwaves could measure the dimensions of the cavity, and thus, reconstruct the change in the size of the cavity. In the end, the cavity became a microphone that could be monitored from far away, allowing the Soviets to eavesdrop on the embassy.

(4) We can measure things that are delicate. When fitting hearing aids, it is very important to know the length of the ear canal because if the hearing aid is too long it can damage the eardrum. So, how do audiologists measure the length of the ear canal? They do not want to stick a ruler in the ear, as, again, the eardrum could be damaged. However, the ear canal is essentially a closed-open air column. By measuring its resonant frequencies, its length can be determined. This is actually one way the ear canal is measured for fitting hearing aids.

(5) We can measure things that are dangerous. As mentioned above, if we have an acoustic cavity, we can learn something about the contents of the cavity. After the Gulf War of 1990–1991, the Americans suspected that chemical weapons might be left in Iraq. So, soldiers would go into bunkers to search for such weapons. However, the artillery shells were not marked as being conventional or chemical weapons. How did the soldiers determine the contents? Obviously, they would not want to open or detonate the shells, as the poison would be released if it were a chemical weapon. So, they conducted an experiment similar to Lab

3.5 to find the resonant frequencies of a cavity. They attached two piezoelectric transducers to the artillery shell. Piezoelectric transducers are simple devices that act as speakers and microphones but work at much higher frequencies than normal acoustic frequencies. They would then scan the frequency going into the speaker from 1 to 1000 kHz and measure the output of the microphone. By analyzing all of the resonant frequencies, they could determine what was inside.

(6) We can measure things inside of us. Medical imaging is a very important application of these techniques. The human heart can have various problems which can be diagnosed if one could look inside. However, it is not an option to screen for problems by opening up the chest, as it would be quite dangerous. Here again, an arrangement of ultrasonic speakers and microphones can produce an image of the heart or many other objects in the body. This is generally known as ultrasound imaging.

All of these examples have one thing in common: they all depend on measuring frequencies and this is collectively known as *spectroscopy*. It is quite amazing how one relatively simple technique can have so many important and even lifesaving applications.

Music of the Spheres—Revisited

Before concluding this chapter, we need to return to where we started—Pythagoras and the Music of the Spheres. We have shown how Pythagoras' simple question "Were does the musical scale come from?" set off a chain of events which led ultimately to the development of modern science and fantastic technological applications. However, we also saw how Pythagoras' work led to the concept of the Music of the Spheres, which held back the field of astronomy for a long time. So, the question is: why didn't the Music of the Spheres work for astronomers?

Pythagoras studied music, and music is based on waves, sound waves, in particular. However, planets are objects and large objects are simply not described by waves. There was no reason for planets to have any connection to music.

Now all of this changed in the early 1900s with the development of quantum mechanics. The physicists who invented quantum mechanics speculated that objects actually *are* waves, although they realized that the larger the object, the harder it would be to see that it is a wave. Even today, we cannot see that anything larger than a molecule has wave-like properties.

However, electrons are very small and generally act as waves. It was originally thought the electrons are small objects that orbit around a positively charged nucleus just like the planets move around the Sun. As it turned out, many properties of atoms and molecules could not be explained with this model.

When it was realized that electrons are waves, physicists changed the model and postulated that the electron wave only has certain modes of vibration, just like the modes of a string or an air column. The modes of a hydrogen atom are three dimensional, but we can make two-dimensional slices through them (Fig. 3.17) and they end up resembling the Chladni plates pictured above.

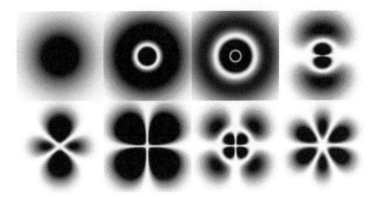

Figure 3.17 Electron modes of a hydrogen atom.

Now, the fact that only certain modes are possible is called "quantization"—leading to the theory of "quantum mechanics". Because of this, the possible modes of the electron are related to integers, just like the musical scale. So, in the end, the motion of an electron in an atom forms an order much like the order of the Music of the Spheres.

In fact, many of the founders of quantum mechanics studied and were influenced by musical acoustics and the problem of consonance and dissonance and this probably helped them to finally understand the true structure of electrons in atoms. This connection was described by a famous physicist Arnold Sommerfeld in the preface to the first edition of *Atomic Structure and Spectral Lines* (1919) this way:

> What we are nowadays hearing of the language of spectra is a true "music of the spheres" within the atom, chords of integral relationships, an order and Harmony that becomes ever more perfect in spite of the manifold variety. All integral laws of spectral lines and atomic theory spring originally from the Quantum Theory.

3.6 Problems

Section 3.1

3.1 The longest bar on the xylophone is 18.30 cm. The bar a major third higher has a length of 16.37. What scale is the xylophone tuned to?

(A) Pentatonic; (B) Pythagorean; (C) Just; (D) Equal temper

3.2 The longest bar on the xylophone is 18.30 cm. How long should the bar a perfect fifth higher be?

(A) 14.94 cm; (B) 25.88 cm; (C) 12.20 cm; (D) 22.41 cm

3.3 I have a metal bar that is 17.3 cm long. The fundamental frequency is 235 Hz. Which of the following frequencies does NOT correspond to a higher mode of vibration of the bar?

(A) 2098.6 Hz; (B) 940.0 Hz; (C) 1269.0 Hz; (D) 648.6 Hz

3.4 A vibrating bar has a fundamental frequency of 440 Hz. Find the frequency of the 3rd overtone and bring it down by octaves until you are in the octave starting with the fundamental. Note, $2.76f$ is the first overtone. What interval does the new note make with the fundamental?

(A) Major third; (B) whole step; (C) half step;
(D) perfect fourth

3.5 I have two identical bars, except that one has a length of 47.2 cm and the other a length of 23.0 cm. What is the ratio of the vibrational frequency of the shorter bar to the longer bar?

3.6 I start with a bar 1.05 m long. How long must I make a second bar to be a perfect fifth above the first?

3.7 A vibrating bar has a fundamental frequency of 416 Hz. Find the frequency of the 3rd overtone and bring it down by octaves until you are in the octave starting with the fundamental. What frequency is this? Note, $2.76f$ is the first overtone

3.8 Assume you have a metal 46 cm long which is tuned to the first note of the scale, "C", 256 Hz. What should the length of a second bar be to make an interval of a perfect fifth higher?

3.9 (a) If a bar is 0.9 m long and has a resonant frequency of 212 Hz, what is the resonant frequency of a bar that is 0.45 m long?

   (b) If you want to raise the resonant frequency by an octave, what length should you use?

3.10 A metal bar is 35 cm long and has a fundamental frequency of 440 Hz. I want to make another bar a perfect fourth *lower* in pitch. How long should I make the new bar?

3.11 A metal bar is 30 cm long and has a fundamental frequency of 480 Hz. I want to make another bar an octave higher in pitch. How long should I make the new bar?

3.12 Say you make a set of wind chimes, which are based on vibrating bars, with certain intervals between them, but then you decide you want them lower in pitch. Can you simply make each bar longer by the same factor and still have the same intervals? This problem will answer that question.

   (a) I happen to find two metal bars, identical except for their length. One is 36 cm long and has a frequency of 550 Hz. The other bar is 32.2 cm long. What is the frequency of the second bar and what is the musical interval between the two bars?

(b) I now make new bars, each one 1.75 times longer than the original bars. What are the frequencies of the new bars?

(c) What is the musical interval between the new bars and is it the same or different from the interval found in part a)?

3.13 (a) I have two identical bars, except that one has a length of 22 cm and the other has a length of 19 cm. What is the interval between them?

(b) How long must a make another bar to be a perfect fifth above the 22 cm bar?

3.14 (a) I happen to find two metal bars, identical except for their length. One is 36 cm long the other is 44.1 cm long. What is the musical interval between them?

(b) I really want an interval of an octave, but, of course, I can only shorten the bars. Which one should I shorten and how long should it be to make the two bars an octave apart?

3.15 A metal bar can have many different modes of vibration. One is a sound wave along the bar:

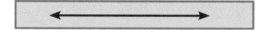

This acts like a closed-closed air column, in that there are antinodes at the ends of the bar. However, the velocity of sound in aluminum, for example, is 5000 m/sec. Sketch the first three modes of vibration of the bar and give the frequency, if the bar is 0.9 m long and made of aluminum.

3.16 A balaphone is an African xylophone. It has wooden bars as the oscillators.

(a) One of the most important things that to know if you want to build a balaphone is exactly where the nodal lines of each bar are, as this is where the bar will be attached to the frame of the instrument. To do this, you can place the bar on a soft piece of foam rubber, sprinkle some sawdust on the bar and keep tapping on the bar.

The sawdust will collect on the nodes of the bar. Sketch the pattern you will see.

(b) Why is this pattern observed, rather than the nodal lines for some other mode of vibration?

(c) From a side view, sketch this mode of vibration.

Section 3.2

3.17 Grand pianos can't be made any bigger because:

(A) They would be dangerous because of the huge amount of tension in the strings.

(B) They would be too hard to play.

(C) They would start to sound too loud.

(D) They would become impossible to tune properly.

(E) They would become too expensive.

3.18 If you are playing a melodic instrument and you change the number of overtones of a note, you will:

(A) Change the tone quality of the note.

(B) Change the pitch of the note.

(C) Lose all sense of pitch.

(D) It is hard to say, because it depends on which overtones are included.

(E) Create a very unpleasant sound.

3.19 If you are playing a non-harmonic or percussion instrument and you change the number of overtones of a note, you will:

(A) Make the pitch sound higher.

(B) Change the pitch of the note.

(C) Lose all sense of pitch.

(D) It is hard to say because most percussion instruments don't have a strong sense of pitch in the first place.

3.20 Why are instruments categorized as melodic or percussion based on their overtone series?

3.21 Series of Frequencies.

State whether the following series correspond to an open-open tube, an open-closed tube, a vibrating bar or none of these. One or more of the lowest frequencies may be missing. If you do recognize the series, give the fundamental frequency

	Type of series	Fundamental
335.2, 502.8, 670.4, 838.0 Hz...		
223.4, 616.6, 1206.4, 1995.0 Hz...		
530.5, 742.7, 954.9, 1167.1 Hz...		
408.1, 734.6, 1754.8, 2120.4 Hz...		

3.22 The overtones of a vibrating metal bar are $1f$, $2.76f$, $5.40f$. What intervals do the second two overtones form with the fundamental? Express the answers as a combination of just intervals which come closest.

3.23 A metal bar is 32 cm long and has a fundamental frequency of 425 Hz.

    (a) What is the frequency of its first overtone?

    (b) I want to make another bar that has a fundamental frequency equal to the first overtone of the original bar. How long should I make the new bar?

Section 3.3

3.24 (a) Sketch individual modes and the nodal lines for an air cavity, closed all sides, in the mode consisting of the second mode in the horizontal direction and the first mode in the vertical direction.

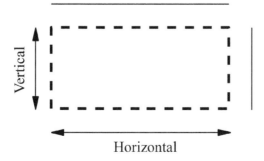

(b) If the horizontal dimension is 0.77 m and the vertical dimension is 42 m, what is the frequency of the mode described in part (a)?

3.25 A rectangular cavity has dimensions 16 cm in the horizontal direction and 34 cm in the vertical direction. If it is in mode number 3 in the horizontal and mode number 1 in the vertical, what is the frequency of this combination mode? Use 343 m/s for the speed of sound and make sure to convert cm to m, if necessary.

3.26 A rectangular cavity has dimensions 14 cm × 23 cm. Find the four lowest frequencies of this cavity. Use 343 m/s for the speed of sound and make sure to convert cm to m, if necessary. Also, be sure to consider both the simple and combination modes.

3.27 A rectangular cavity filled with air has resonant frequencies at 500 Hz, 700 Hz, and 860 Hz. If the speed of sound in carbon dioxide is 1.25 times smaller, what would these frequencies become if the cavity was filled with carbon dioxide? (Start with $f \cdot \lambda = v$).

3.28 What are the first four frequencies of a square cavity 20 cm on each side?

3.29 A 2D cavity has dimensions 30 by 40 cm. Fill in the following table and identify all of the intervals with the fundamental frequency. If possible, identify the chord that these intervals form. (One note will not fit.)

	$f_1 =$	$f_2 =$
$g_1 =$	$(f_1, g_1) =$	$(f_2, g_1) =$
$g_2 =$	$(f_1, g_2) =$	$(f_2, g_2) =$

3.30 (a) Determine whether the following series of frequencies come from a closed-closed air column, and open-closed air column, or neither. The lowest frequencies may not have been recorded. If it is a proper series given the fundamental frequency

	Type of series	Fundamental
465.6, 776.0, 1086.4, 1396.8 Hz...		
150.3, 310.1, 480.9, 660.4 Hz...		
545.1, 726.8, 908.5, 1090.2 Hz...		

(b) The following series correspond to a closed rectangular cavity, except the lowest frequency might be missing. Identify the mode corresponding to each frequency. Let *f* be the series with the lowest fundamental frequency

Series 1	350.0 Hz	425.0 Hz	550.6 Hz	700.0 Hz	818.9 Hz	850.0 Hz
Mode						

Series 2	550.0 Hz	630.0 Hz	687.4 Hz	825.0 Hz	836.3 Hz	1038.0 Hz
Mode						

3.31 Suppose we measure the following frequencies for a rectangular cavity. Figure out the dimensions of the cavity, assuming the speed of sound is 343 m/sec. You will need to identify the two regular series. To prove that you did this correctly, show how to make up the rest of the combination frequencies. There was one problem with the apparatus, however; frequencies lower than 400 Hz could not be measured. Thus, the fundamental of one series is missing

Series	424.7	555.7	716.6	833.0	849.4	921.9	1074.9	1111.3	1155.8	1274.1	1323.5	1370.0	1433.2
Mode													

3.32 (a) Sketch the nodal lines for a closed rectangular air cavity in the mode consisting of the second mode in the long direction and the third mode in the short direction.

(b) If the cavity is 0.25 m by 0.15 m, what is the frequency of this mode?

3.33 If we place a microphone in the circle labeled (1), find the frequencies of the lowest three modes that a microphone would detect. The dimensions of the cavity are 18 cm by 10 cm. Hint, you must think about where the nodal lines are. Repeat this for the case where the microphone is put in the circle labeled (2). Circle (2) is exactly 1/4 of the way in from the side. Use 343 m/s for the speed of sound and make sure to convert cm to m, if necessary. Also, be sure to consider both the simple and combination modes.

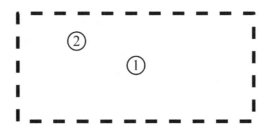

3.34 (a) A rectangular cavity 30 cm by 10 cm has a microphone in the bottom plate at position (1). What are the first three modes of the closed cavity that the microphone can detect? (Hint, if made has a nodal line at the position of the microphone, the microphone will not detect that mode).

(b) Repeat (a) for the microphone at position (2).

(c) Repeat (a) for the microphone at position (3).

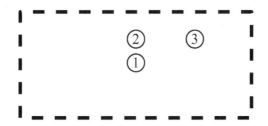

3.35 In analyzing the waves in the rectangular cavities, we must consider the fact that the waves are two-dimensional instead of one-dimensional in an air column. The question then comes up: why don't we have to worry about the third dimension? Assume we have a rectangular cavity that is 30 cm in length, 18 cm in width, and 1 cm in height, and we can only measure frequencies up to 10,000 Hz.

(a) What is the frequency of the fundamental mode in each direction—length (f), width (g), and height (h)?

(b) In addition to the 2D modes, we can have simple modes in the vertical direction and combination modes given by $F = \sqrt{f^2 + g^2 + h^2}$. Why do we not have to worry about this possibility?

Section 3.4

3.36 Why do modern physics and quantum mechanics have more in common with music than classical physics?

3.37 So far, we have analyzed a tube or rectangular cavity and determined the resonant frequencies of the cavity. Really, what is useful is the reverse. If we measure the frequencies, we can determine the dimension of the cavity. This will be true even if we cannot see the cavity directly (either it is too small or too far away).

(a) What length of tube will produce the following set of frequencies: 234 Hz, 468 Hz, 702 Hz, ...?

(b) What length of tube will produce the following set of frequencies: 986 Hz, 1479 Hz, 1972 Hz, ...? The difference here is that I did not record the fundamental, so you will need to figure that out. Assume in both parts that the speed of sound is 343 m/sec.

Section 3.5

3.38 Assume the following diagram gives you the frequencies of the resonant modes of a rectangular cavity as you change the length of one of the sides:

(a) Label the line with "g_1" which represents the fundamental mode in the direction that does NOT change.

(b) Label the line "f_2" which represents the second mode in the direction that does change.

(c) Sketch the line which would represent the combination frequency, $\sqrt{f_2^2 + g_1^2}$, of the frequencies given in part (a) and (b).

(d) At what length is $f_2 = g_1$?

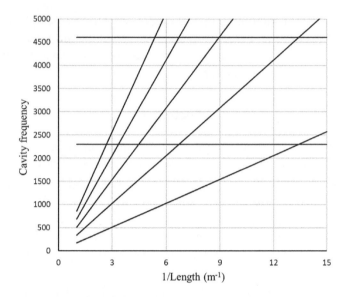

3.39 Cavity Mode Diagrams

Once again we have a rectangular cavity in which one side is kept fixed while the other side is varied. We plot the resonant frequencies as a function of $1/L$ were L is the length that we are changing:

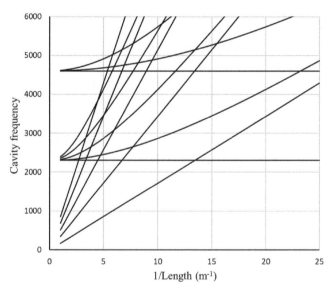

If f_1, f_2, f_3 ... represent the series of regular resonances in the direction we are changing and g_1, g_2, g_3 ... represent the series of regular resonances in the fixed direction, answer the following:

(a) Over what range of lengths would the order of resonances going from low frequency to high frequency be: $f_1, g_1,$ $\sqrt{f_1^2 + g_1^2},...$

(b) Over what range of lengths would the order of resonances going from low frequency to high frequency be: $f_1, f_2, g_1,$ $\sqrt{f_1^2 + g_1^2},...$

(c) What is the ordering of the first four frequencies at a length of 0.05 m?

(d) At what length is $g_2, \sqrt{f_1^2 + g_1^2}$?

3.40 Cavity Mode Diagrams

Using the graph in Problem 3.39, answer the following questions:

(a) Label f_1, f_2, g_1, g_2.

(b) Over what range of lengths would the order of resonances going from low frequency to high frequency be: $g_1, f_1, \sqrt{f_1^2 + g_1^2},...$

(c) What is the ordering of the first four frequencies at a length of 0.2 m?

(d) At what length is $f_2 = \sqrt{f_1^2 + g_1^2}$?

Chapter 4

Waves and Interference

4.1   General Properties of Waves

Now that we have learned quite a bit about waves and how they relate to music and musical instruments, it is useful to go back and discuss waves from a more general, and more technical, perspective. Many of the concepts will not come as a surprise, as you have begun to think about waves more carefully.

Waves in General

1. Waves require a medium.

    Perhaps one of the most unusual features of waves, as compared to objects, is that waves cannot exist on their own. They require a medium to travel in. For example, sound waves need air. If you place an alarm clock in a large jar and pump out all of the air, you cannot hear the alarm clock ring. The sound waves simply do not exist in a vacuum. The waves in the ocean only exist because of the water. The waves are not objects in themselves; they don't exist without the medium. In fact, waves represent a *disturbance* of the medium.

2. A medium that supports waves must have a stable equilibrium.

    An equilibrium is a state of any system that does not change over time. For example, a perfectly flat pond is in a

The Music of Physics: An Introduction to the Harmonies of Nature
George N. Gibson
Copyright © 2024 Jenny Stanford Publishing Pte. Ltd.
ISBN 978-981-4968-99-7 (Hardcover), 978-1-003-48481-3 (eBook)
www.jennystanford.com

state of equilibrium and no waves are present on the pond. However, if we throw a rock into the pond, the water (or the medium) is disturbed, creating ripples around the point that the rock hit the water. The ripples are *not* static and move through the water. It is this disturbance of the medium that *is* the wave. There are two kinds of equilibrium: *stable* and *unstable*. If we push a system way from equilibrium and it tends to return to its equilibrium, it is said to be stable. For example, let's consider a marble at the bottom of a bowl (Fig. 4.1).

Figure 4.1 An example of a stable equilibrium.

While the marble sits at the bottom of the bowl, it is in equilibrium–it will not move. If we displace the marble slightly from the bottom, the marble will tend to roll back to its equilibrium point. This is the defining characteristic of a *stable* equilibrium. Of course, it will overshoot this point and roll up the other side of the bowl. But, now, it will still tend to come back towards the equilibrium and will end up oscillating back and forth around the equilibrium point. This ability to oscillate, as you know, is a critical property of a wave.

Now consider an inverted bowl with a marble on top (Fig. 4.2).

Figure 4.2 An example of an unstable equilibrium.

In this case, the marble will again sit on the top of the bowl without moving. However, if we displace it slightly, it will tend to just roll off the bowl altogether. It will certainly not oscillate or even try to come back. This is an *unstable* equilibrium. A medium with an unstable equilibrium will not support a wave.

3. Restoring force.

   A stable equilibrium exists because there is a *restoring force* returning the medium to its equilibrium. The restoring force always points back to the equilibrium point and is a property of the medium. For sound waves, the medium is air and the restoring force is air pressure. For water waves, the medium is the water and the restoring force is either surface tension or gravity.

4. A wave, or a disturbance of a medium, does not stay localized.

   A wave spreads out in one, two, or three dimensions. This aspect is also quite different from normal objects. An object like a baseball generally stays together. If you place it on a table, it just sits there. In contrast, a wave is always in motion—sound waves or water waves do not just stand still. Also, waves tend to spread out. If you throw a baseball, the baseball arrives as a whole at some other location; it does not spread out over a large area. Waves do not stay together; they spread out. If sound waves stayed together like an object, I could only lecture to one person at a time. The sound waves that I create as I talk would only get delivered to one location. In fact, waves spread out, so that the sound waves I produce fill the entire room and everyone in the room can hear me.

5. A wave, or a disturbance of a medium, carries energy.

   This is a very important point. If waves do not exist on their own and are only a disturbance of a medium, one might wonder if they really have a separate existence. The reason we know that waves exist is that they carry energy. It takes energy to move something from its equilibrium. This is the initial disturbance. If the disturbance moves, so must the energy. Therefore, the energy moves along with the wave. If waves did not carry energy, you could not hear me speak. I create sound waves, and the waves travel to your ear and make the eardrum vibrate. However, it takes energy to make the eardrum move, so the sound wave must carry energy. A more dramatic example of waves carrying energy is a tsunami. An earthquake in Japan can produce an enormous wave in the ocean called a tsunami. The tsunami can travel across the entire Pacific Ocean and crash on the North American coast and destroy a town. This represents a huge amount of energy traveling a very long distance. Note that it is the disturbance which travels across the ocean. The water where the tsunami was created does not travel very far. Thus, it should be clear that waves have a physical existence, just like objects, and waves simply follow different physical laws than objects, and we are trying to understand these laws.

6. Longitudinal and transverse waves.

There are actually two very distinct kinds of waves or disturbances, depending on the medium. The type of wave has to do with the direction in which the medium moves as compared to the direction in which the wave moves. Consider a wave on a string. The wave moves along the string, but the string, itself, moves up and down or back and forth. In this situation, the medium (the string) is moving in a direction perpendicular to the direction that the wave moves. This is called a *transverse* wave.

Sound waves are rather different. Looking back at Section 3.6, we considered sound waves traveling up and down a tube. Now think about what the individual air molecules are doing. They are vibrating back and forth along the tube, as well. So, in this case, the medium is moving in the same direction as the wave. This is called a *longitudinal* wave. These two possibilities are shown in Fig. 4.3.

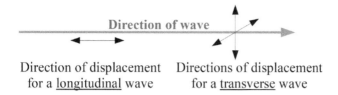

Direction of displacement
for a <u>longitudinal</u> wave

Directions of displacement
for a <u>transverse</u> wave

Figure 4.3 Diagram showing the difference between longitudinal and transverse waves.

Notice that for transverse waves, we mentioned two possible directions for the string: up and down and back and forth. This is true of all transverse waves: there are always two directions for the medium to move that are perpendicular to the direction of the wave. The particular direction that the medium moves is referred to as the *polarization* of the wave. Transverse waves have two possible polarizations. The idea of polarization is quite important in discussing the properties of light—a kind of wave that we will explore later in this chapter.

For longitudinal waves, the medium has only one direction that it can move—that is parallel to the direction of the wave. So, in this case, there is no need for the concept of polarization.

7. Wave velocity

We have already discussed the velocity of a wave. As mentioned above, this is a curious property of a wave—it is always in motion and always travels with the same speed, the wave velocity. The velocity of a wave depends on the properties of the medium in a rather simple way:

$$v = \sqrt{\frac{\text{restoring force}}{\text{density of medium}}} = \frac{\text{length}}{\text{time}}. \tag{4.1}$$

We discussed the restoring force above—that is the force tending to return the medium to its equilibrium. The density of the medium is just what it sounds like. Waves in a thick liquid, like molasses, will travel more slowly than waves in a thin liquid, like water (assuming, of course, that the restoring force is the same). The units on restoring forces and densities are a bit complicated, but the combination of the square root of the ratio of the two does work out to length/time, as it should.

So, the speed of sound is given by:

$$v_{sound} = \sqrt{\frac{\text{pressure}}{\text{density of gas}}}. \tag{4.2}$$

Technically, this is true as long as this is evaluated adiabatically. However, this is a somewhat technical term and is beyond the scope of this course. Nevertheless, the dependence of the velocity of sound on pressure and density in Eq. 4.2 is correct.

Similarly, the speed of a wave on a string is given by:

$$v_{string} = \sqrt{\frac{\text{tension}}{\text{density of string}}} \tag{4.3}$$

The restoring force for a string is the tension on the string. Looking back at Section 2.5, we can now see in a more general way where the expression for the velocity of a wave on a string came from.

8. Waves are not solid.

Perhaps the most unusual aspect of waves is that they can move through each other without affecting the other wave. This is very different from solid objects. If two solid objects try to pass through each other, either they bounce off each other or one or both objects break up. If waves bounced off of each other, conversations would be very difficult. It would be possible to block a sound wave with another sound wave. This does not happen.

9. Superposition and interference.

Based on the previous point, it can happen that two waves are at the same place at the same time. How do we describe this situation? Instead of displacing each other or bouncing off each other, the waves simply add together. This is called the superposition of waves. One wave is superimposed on the other. Although this statement is easy to make—that two waves in the same space at the same time simply add together—actually calculating the effects of this can be quite tricky. A number of unusual things can happen because of this property of waves and are generally called interference effects. We will explore several consequences of superposition in the next few sections.

4.2 Constructive and Destructive Interference

In the last section, we discussed the fact that waves can move through each other, which means that they can be in the same place at the same time. This is very different from solid objects. Thus, we need to know how to handle this situation. As it turns out, when waves are at the same place at the same time, the amplitudes of the waves simply add together and this is really all we need to know! However, the consequences of this are profound and sometimes startling.

To start exploring the implications of the statement above, let's consider two waves with the same frequency traveling in the same direction (Fig. 4.4).

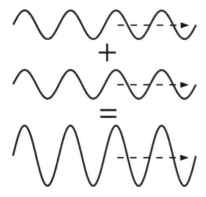

Figure 4.4 Constructive interference.

If we add these two waves together, point-by-point, we end up with a new wave that looks pretty much like the original waves but its amplitude is larger. This situation, where the resulting wave is bigger than either of the two original, is called *constructive interference*. The waves are adding together to form a bigger wave. You may be thinking that this is pretty obvious and natural—of course, the sum of two waves will be bigger than each wave on its own. However, carefully consider the next situation, again where two waves with the same frequency are traveling in the same direction (Fig. 4.5).

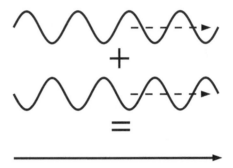

Figure 4.5 Destructive interference.

Now what happens if we add these waves together? When the first wave is up, the second wave is down and the two add to zero. When the first wave is down and the second is up, they again add to zero. In fact, at all points the two waves exactly cancel each other out and there is no wave left. This is the single most amazing aspect of waves. The sum of two waves can be less than either wave, alone, and can even be zero. This is called *destructive interference*.

When the peaks of the waves line up, there is constructive interference. Often, this is described by saying the waves are "in phase". Although this phrase is not so important for this course, it is so commonly used that you may hear it in other settings. Similarly, when the peaks of one wave line up with the valleys of the other, the waves are said to be "out of phase". Phase, itself, is an important aspect of waves, but we will not use this concept in this course.

How could we observe this difference between constructive and destructive interference? Given the fact that in one case we get a bigger (or louder) wave, and in the other case we get nothing, there should be a pretty big difference between the two. We will explore how to hear this difference in detail in Lab 4.1.

The most important requirement for interference is to have at least two waves. One wave alone does not show interference. We shall see that there are many ways to create a pair of waves to produce interference. The simplest way to create two sound waves is to use two speakers. If we place them side-by-side, point them in the same direction and play the same frequency in each, we have just the situation described above to produce constructive interference (Fig. 4.6).

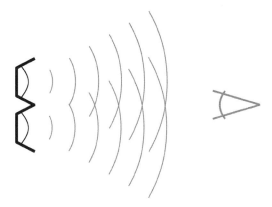

Figure 4.6 Constructive interference from two speakers.

If we stand in front of the two speakers, we will hear a tone louder than the individual speakers would produce. The two waves are in phase. Now imagine that we start moving on of the speakers back (Fig. 4.7).

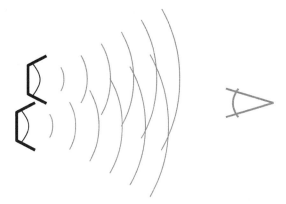

Figure 4.7 Destructive interference from two speakers.

At some point, the two waves will be out of phase—that is, the peaks of one line up with the valleys of the other creating the conditions for destructive interference. If we stand in front of the speakers right now, we will not hear anything. This must be experienced to really appreciate. Equally as strange, if you now block one speaker, the destructive interference goes away and you hear the unblocked speaker. In other words, the sound gets louder as you block one speaker.

How far back must we move the speaker to go from constructive to destructive interference? We know that the distance between peaks in a wave is equal to the wavelength. If we look back at Fig. 4.5 we see that the waves are shifted by half of a wavelength. So, in the example with the speakers, we must move the speaker back by one half of a wavelength.

What happens if we keep moving the speaker back? At some point the peaks of the two waves will again line up (Fig. 4.8).

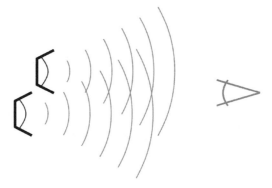

Figure 4.8 Constructive interference where the speakers are not side-by-side.

At this position, we will again have constructive interference. Here, the waves have been shifted by a full wavelength and the peaks line up again, like in Fig. 4.4. As the speaker is moved back the waves alternate between constructive and destructive interference.

What the example of the speakers shows is that it is the separation of the two speakers that determines whether there will be constructive or destructive interference. If the speakers are at the same position, there will be constructive interference at all points directly in front of the speaker. If the speakers are separated by half a wavelength, then there is destructive interference, regardless of how far or close you are to the speakers.

However, it already has become apparent that this is not the whole story, because if you keep moving the speaker you again can achieve constructive interference. This can be fairly easily incorporated into our picture by saying that if the separation of the speakers in a multiple of a wavelength then there will be constructive interference. Note that zero separation can always be considered a multiple of a wavelength.

What about destructive interference? We know that if the speakers are separated by half a wavelength there is destructive interference. However, if we move an additional full wavelength, we will still have destructive interference. So, this case is a bit harder to state, but if the separation is equal to half a wavelength plus a multiple of a wavelength, there will be destructive interference. Thus, we have described the conditions under which we will have constructive and destructive interference for two waves with the same frequency traveling in the same direction. Unfortunately, the conditions have been expressed in a cumbersome way that is not easily applied to more complex situations. So, before going on to other examples, we need a more mathematically concise way of stating the conditions for constructive and destructive interference.

The proper way to define the conditions for having constructive or destructive interference requires knowing the distance from the observation point to the source of each of the two waves. Since there must be two waves for interference to occur, there are also two distances involved, R_1 and R_2. For two waves traveling in the same direction, these two distances are shown in Fig. 4.9.

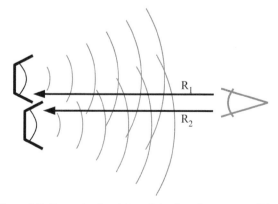

Figure 4.9 Geometry for determining interference condition.

When we discussed interference above, it became apparent that it is the separation between the two speakers that determined whether the interference was constructive or destructive. From this diagram, we see that the separation is given by $R_1 - R_2$. So, really, *it is the difference in path length from each source to the observer that determines whether the interference is constructive or destructive.* I emphasize this point, because it is true in *all* situations involving interference. The only difficulty lies in properly applying this concept.

With this more rigorous statement about interference, we can now write down mathematically the conditions for interference.

Constructive Interference

We saw that when the two speakers are right next to each other, we have constructive interference. However, if the speakers are next to each other, the distance from each to the observer must be the same, which means that $R_1 = R_2$. Or, we can write that $R_1 - R_2 = 0$. That is, the path difference must be equal to zero. But, we also saw that if we move one speaker by a whole wavelength, we still have constructive interference. Remember that we use the Greek letter λ for wavelength. With this, our condition for constructive interference can be written:

$$|R_1 - R_2| = 0 + n\lambda. \tag{4.4}$$

Here, the variable n is used to specify an integer and can take on any value, as long as it is an integer, including zero. This ensures that we only add whole numbers of wavelengths. I also added an absolute value around the path difference, as it does not matter which is longer or shorter, only the magnitude of the difference.

Destructive Interference

Once we have the condition for constructive interference, destructive interference is a straightforward extension. The basic requirement for destructive interference is that the two waves are shifted by half a wavelength. This means that the path difference for the two waves must be: $R_1 - R_2 = \lambda/2$. But, since we can always shift a wave by one full wavelength, the full condition for destructive interference becomes:

$$|R_1 - R_2| = \lambda/2 + n\lambda. \tag{4.5}$$

Now that we have mathematical statements for the requirements for constructive and destructive interference, we can apply them to a new situation and see what happens.

Waves with the Same Frequency Traveling in Opposite Directions

To create two waves traveling in opposite directions, we can take our two speakers and point them at each other, as shown in Fig. 4.10. We again want to find the conditions for constructive and destructive interference. As we have seen, the simplest way to get constructive interference is for the distance from the observer to each source to be equal. Using our mathematical terminology, we want $R_1 - R_2 = 0$, or $R_1 = R_2$. Looking at the figure above, we see that the point where the two paths are equal is exactly midway between the two speakers (point M in Fig. 4.10). At this point, there will be constructive interference, and the sound will be strong.

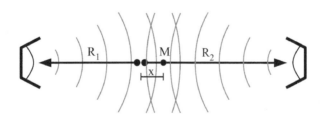

Figure 4.10 Interference of waves traveling in opposite directions.

It makes sense to use the midpoint as a reference, as we know that we have constructive interference. How far must we move our observer to get to destructive interference? If we move to the left by an amount x, the distance R_1 decreases by x and the distance R_2 increases by x. If R_1 decreases and R_2 increases, the difference between the two $|R_1 - R_2|$ increases by an amount $2x$. So, at the point x, the path difference is $|R_1 - R_2| = 2x$.

Now comes the tricky part. If $2x$ happens to be equal to $\lambda/2$, we have met the conditions for destructive interference. Therefore, if $2x = \lambda/2$, or $x = \lambda/4$, we have destructive interference. To put it another way, in the situation above, if you move one quarter of a wavelength away from the midpoint, you will find destructive interference and the sound will sound very weak, or you might not hear anything at all.

What happens if we keep moving our observation point? If the path difference, $2x$, equal one whole wavelength, we will have constructive interference, $2x = \lambda$. Notice that we did not change $|R_1 - R_2|$, just the interference condition, going from $\lambda/2$ to λ. Solving for x, we have $x = \lambda/2$. In other words, if we move by half a wavelength, we will again have constructive interference and the sound will be loud.

As we keep moving the observation point, we will find that we keep going through points of constructive and destructive interference. This is a bit more complicated than the first example, where we had either constructive or destructive interference regardless of where we listened. In this case, whether there is constructive or destructive interference depends on where we are listening. However, the fundamental conditions on the path difference are still the same.

What does this pattern of constructive and destructive interference look like? We can map it out by indicating where we have constructive (x) and destructive ($\bullet$) interference in Fig. 4.11.

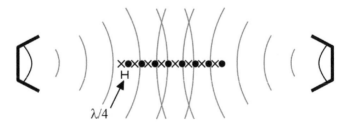

Figure 4.11 Standing wave formed from traveling waves.

What we see is a repeating pattern of constructive and destructive interference, and it takes a distance of $\lambda/4$ to get from one to the other. Where have we seen this pattern before? At a point of constructive interference, the amplitude of the wave is large and this is just like an antinode. At a point of destructive interference, the amplitude is zero and this is like a node. So, if we think of the points above as antinodes and nodes, we see that we have exactly the same pattern of nodes and antinodes as in a standing wave (see Fig. 2.5, for example). From this, we must conclude that two waves traveling in opposite directions create a standing wave with the same frequency. It is quite common in physics to have two different trains of thought come together and reveal a deeper connection between the two. This is similar to the realization that the shape of a wave is directly linked to the overtones present and both are responsible for different timbres. We will see another such connection in Chapter 5.

Translating the interference conditions into mathematical statements is an essential part of physics and can be quite difficult at first. Moreover, a rather subtle connection was made that you might not have noticed. On the one hand, we have some physical situation or geometry. This refers to the placement of the speakers and the position of the observer. This really has nothing to do with waves and it simply depends on how the problem was set up. However, given a particular setup, you can always figure out the path length from the observer to the two sources of the waves that are going to interference and hence you can also find the path difference $R_1 - R_2$. Again, this is separate from waves or interference.

Completely independent of the geometry of the setup, there is a property of waves called superposition that can lead to constructive or destructive interference. We expressed these conditions mathematically in Eqs. 4.4 and 4.5. But, again, in these equations, $|R_1 - R_2|$ is determined from the geometry of the problem. These two aspects must be understood separately: how to calculate the path difference and how to impose the conditions which determine the type of interference.

Only then should these two aspects be combined to determine whether there is constructive or destructive interference at a particular location of the observer. The link between the two aspects is $|R_1 - R_2|$: it is determined from the geometry and then compared to the wavelength to find the interference condition. This can be summarized in a diagram, below, using waves traveling in opposite directions as an example.

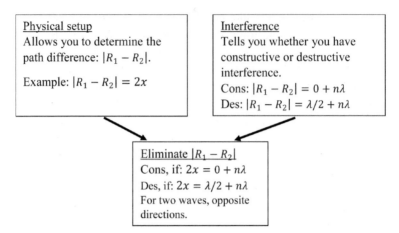

Physical setup
Allows you to determine the
path difference: $|R_1 - R_2|$.

Example: $|R_1 - R_2| = 2x$

Interference
Tells you whether you have
constructive or destructive
interference.
Cons: $|R_1 - R_2| = 0 + n\lambda$
Des: $|R_1 - R_2| = \lambda/2 + n\lambda$

Eliminate $|R_1 - R_2|$
Cons, if: $2x = 0 + n\lambda$
Des, if: $2x = \lambda/2 + n\lambda$
For two waves, opposite
directions.

In the next sections, we will explore many more situations for seeing constructive and destructive interference.

4.3 Interference in Reflections

Another kind of interference that is rather easy to observe is interference though reflection. In the previous section, we considered situations where there are two sources of waves, as there must be at least two waves to have interference. However, there are times when one wave can produce two waves and those two waves then interfere.

Imagine the cross-section of a standard wall. There is an exterior wall, an interior wall and a gap in between for insulation, as in Fig. 4.12.

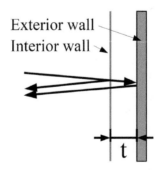

Exterior wall
Interior wall

t

Figure 4.12 Interference of sound waves through reflection.

When a sound wave strikes this structure, it produces two reflected waves that can interfere. The question is, for a particular frequency, will there be constructive or destructive interference?

Using the method discussed in the previous section, there are always two parts to a question like this. The first part involves using the figure to determine the difference in path length between the observer and the two waves.

Looking back at Fig. 4.12, we see that one of the waves reflects off of the first wall and travels straight back to the observer. The second wave travels through to the second wall before reflecting back. If the thickness of the gap is called "t", then the second wave travels a distance $2t$ compared to the first wave. Remember it has to travel through the gap in both directions. So, the difference in path length traveled by the two waves, $|R_2 - R_1|$, is just $2t$.

The second step in the analysis is recalling the conditions for constructive and destructive interference, shown in Eqs. 4.4 and 4.5 above. In this example, we can ask what frequencies or wavelengths are suppressed due to destructive interference. The simplest condition for destructive interference is $|R_2 - R_1| = 1/2$. We can now bring these two pieces of information together:

Path length difference from diagram: $|R_2 - R_1| = 2t$.

Path length difference for destructive interference: $|R_2 - R_1| = \lambda/2$.

Then, for destructive interference, we have $2t = \lambda/2$, or $\lambda = 4t$. For constructive interference, the condition will be $\lambda = 2t$.

Another example of interference of sound comes from reflection from steps. A common situation involves hitting a bass drum in a football field in front of a set of bleachers, which we already encountered in Fig. 1.1. As before, if the depth of each step is given by "D", then the conditions for destructive interference are $D = \lambda/4$. There will be constructive interference if $D = \lambda/2$. The bass drum produces many frequencies and the frequencies that produce constructive interference will reflect strongly. So, even though the bass drum does not produce a distinct pitch, the reflected sound does have a clear pitch corresponding to the frequencies that interfere constructively. This turns out to be the same pitch that we concluded in Chapter 1 from Fig. 1.1. This is another example of the same physical result being understood in different ways. In Chapter 1 we found the pitch by considering the repetition of the echoes in time, while in this chapter, we considered which frequencies experience constructive interference.

Worked Example

Two speakers are arranged in a room as shown in the diagram. The speakers are playing a pure tone with a wavelength of 20 cm. At each location, determine if there will be constructive interference, destructive interference, something in between, or cannot determine.

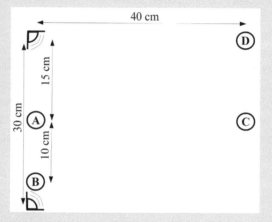

A is the same distance from each speaker, so $|R_1 - R_2| = 0$, so there is constructive interference.

B is 25 cm from one speaker and 5 cm from the other, so $|R_1 - R_2| = (25-5)$ cm $= 20$ cm. This is one wavelength λ, so there is constructive interference.

C is on the centerline between the speakers, so, as for A, $R_1 = R_2$ and there is constructive interference.

D is 40 cm from one speaker. To determine the distance to the other speaker, we need the Pythagorean Theorem. The distance will be square root of $(40 \text{ cm})^2 + (30 \text{ cm})^2 = 50$ cm. So, in this case $|R_1 - R_2| = (50-40)$ cm $= 10$ cm $= \lambda/2$, so there is destructive interference.

4.4   Interference Patterns

In the previous section, we considered two geometries for the interference of two waves with the same frequency. In the first case, the waves were traveling in the same direction, and in the second, the waves were traveling at each other in opposite directions. In both cases, we considered whether we would encounter constructive or destructive interference as we moved along the same line that the waves were moving. This reduced the problem to one dimension, just as we began our discussion of waves in Chapter 3. Although one dimension is easier to understand, many important effects happen in two or three dimensions, so in this section, we will consider a two-dimensional geometry. Under these conditions, the points of constructive and destructive interference form a complex pattern, which we will refer to as an interference pattern.

To begin the discussion, consider two sources of sound that give off waves in all directions, Fig. 4.13. One source and its wave is colored black, while the other is grey. Initially, the sources are right on top of each other, so you can't yet see the grey source:

Figure 4.13 Constructive interference of two sources on top of each other.

Since the sources are right on top of each other, wherever you stand, you will be the same distance from each source. Thus, the path difference to each source will be zero and there will be constructive interference everywhere.

Now, let's consider what happens if we move one speaker by half a wavelength in the vertical direction, as shown in Fig. 4.14.

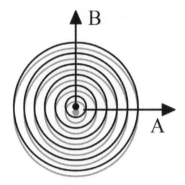

Figure 4.14 Destructive interference occurring in the vertical direction.

Things begin to look more complicated. If we look along direction A, we find that no matter where we are on that line, we are always the same distance from each speaker. Therefore, we will find constructive interference. So, even though we moved one of the speakers, we still have constructive interference along A. This is not true for direction B. Along this line, the speakers are displaced from each other by half a wavelength. In the last section, we found that under these conditions, the waves will interfere destructively, and we will not hear any sound in this direction. We can imagine ourselves walking in a big circle around the two speakers and marking the points where we find constructive (x) and destructive (•) interference, shown in Fig. 4.15.

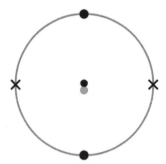

Figure 4.15 Alternative view of the interference pattern from Fig. 4.14.

Of course, we can keep going. Let's move the speakers apart by a full wavelength, as shown in Fig. 4.16.

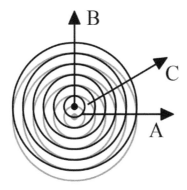

Figure 4.16 Interference pattern with sources one wavelength apart.

We will analyze this qualitatively. How many points of constructive and destructive interference will you find as you move around the circle? Again, nothing has changed in direction A. We will always be equidistant from each speaker and get constructive interference. Along direction B, the speakers are now displaced by a full wavelength also giving constructive interference. So along the direction that the speaker is displaced, we go repeated through constructive and destructive interference. Now, in the present case, we have constructive interference at both A and B. Does this mean that we will have constructive interference at all points in between A and B? The answer is definitely no. At A, the path difference is zero while at B the path difference is a wavelength, or λ. As we move from A to B the path difference will change continuously from zero to λ, so there must be some point where the path difference is exactly $\lambda/2$. At this point, we will have destructive interference. This is marked in the figure above by C. The interference pattern along a circle surrounding the speakers now looks like Fig. 4.17.

And, of course, we can keep moving the speakers apart. Every time we move a half of a wavelength, the point at the top of the figure will go from constructive to destructive interference and a new point of constructive interference will appear on the circle, OR the point at the top of the figure will go from destructive to constructive interference and a new point of destructive interference will appear on the circle.

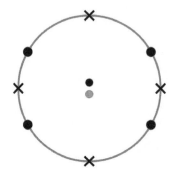

Figure 4.17 Alternative view of interference pattern from Fig. 4.16.

As the separation increases, the pattern will become increasingly complicated, but in principle, we can figure out the pattern of constructive and destructive interference if we know the separation of the two sources.

Although the interference pattern can become quite complicated, there is one fairly simple aspect to it and which is also the most useful. So far, we considered the interference pattern as though we were walking in a big circle around the two sources of waves. Now let's restrict our attention to a smaller region where we are pretty much right in front of the two sources, Fig. 4.18.

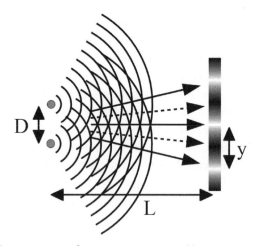

Figure 4.18 Interference pattern created by two sources.

In this picture, there are two sources of waves separated by a distance D. If you look a distance L away from the two sources, you will find points of constructive and destructive interference (shown by the grey-scale pattern, constructive is white and destructive is black). The solid arrows show where there is constructive interference, while the dashed arrows show destructive interference. In turns out that there is a rather simple expression giving the spacing between the points of constructive interference (or between points of destructive interference):

$$y = \frac{\lambda D}{L}, \qquad (4.6)$$

where λ is the wavelength of the wave.

This effect is important for many reasons. This particular interference pattern is fairly simple to setup and observe. One just places two sources close together and moves back a distance L. If the wavelength is known, one can measure the separation D. If the wavelength is very small, one can measure very small separations this way. Alternatively, if the separation is known, one can make a measurement of the wavelength of a wave. The interference pattern and Eq. 4.6 can be verified with Lab 4.2.

However, even more important than these applications is the fact that if you do see an interference pattern, then you know that you are dealing with waves. How is this important? Consider the photograph shown in Fig. 4.19.

The exposure is longer and longer going from (a) to (d). In (d) it is quite clear that there is an interference pattern. What makes this result so remarkable is that the source in this case was a beam of electrons. Now, electrons are particles that have a mass, like a rock or a baseball. Until the early 1900s, no one ever imagined that the matter that we are made of could be a wave. Many types of waves were understood, such as sound and light waves as well as the phenomenon of interference. So, when an interference pattern was made with electrons it proved that electrons, at least sometimes, acted as waves. This completely changed the course of physics and led to quantum mechanics and much of the modern technology that we now take for granted.

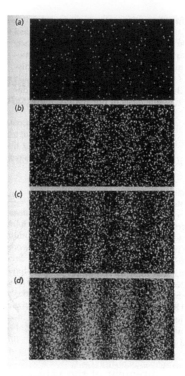

Figure 4.19 Interference pattern created by single electrons.

The actual situation is even more complicated. Although picture (d) shows an interference pattern, all of the photographs are made up of small dots. Each dot corresponds to an electron hitting the photographic film. This indicates that electrons sometimes do act as particles, but only when they are detected. While undetected they travel as waves. This dual nature of small particles is very hard to comprehend and, yet, it has been verified by thousands of experiments.

One question is worth asking: why did it take so long to realize the wave nature of matter? We can get some idea by looking at the expression for the spacing of the bands in the interference pattern (Eq. 4.6). It just so happens that the wavelength of the waves that matter is composed of is very small. This makes the spacing, y, also very small and, thus, hard to see. The interference pattern is very fine and easily blurred out. If you cannot make out the constructive and destructive interference, then you would not think that you are observing a wave. So, it took physicists some time to figure out how to actually see the interference pattern.

The wavelength of sound waves in much greater than the wavelength of matter waves, so the interference of sound is much easier to observe, and indeed it is rather straightforward to set up an experiment to detect this interference (Lab 4.1). As sound bounces around a room, it forms a very complex interference pattern that is hard to predict but can lead to spots in the room that are loud due to constructive interference and spots that are soft due to destructive interference. Moreover, each frequency of sound has a different wavelength, and so each frequency will have a different interference pattern making the sound characteristics change significantly from point to point. If you have ever watched an accomplished musician play, they often seem to sway, moving their heads quite a bit. It is thought that this is an instinctive response to the complex interference pattern: by moving one's head, one can "average out" this interference pattern leading to a more accurate perception of the sound being produced.

4.5 Interference in Time: Beats

In the last three sections, we have been considering various possibilities for the interference of two waves with the same frequency. This interference gives rise to a number of important and interesting effects. In this section, we explore a different situation: what happens when two waves with different frequencies interfere?

Although it may seem that this will make the problem more complicated, it actually leads to a rather simple result. To be begin the discussion, let's consider two waves with different frequencies traveling in the same direction (Fig. 4.20).

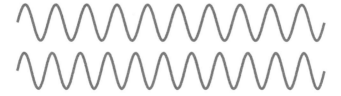

Figure 4.20 Two waves with slightly different frequencies shown in time.

The top wave has a frequency of 10 Hz, while the bottom wave has a frequency of 11 Hz. One second of each wave is plotted, so the top wave shows 10 oscillations and the bottom wave, 11 oscillations.

At the start, the two waves are in phase; that is, the peaks and valleys line up. At this point they add together constructively. In Section 4.2, we saw that two waves with the same frequency traveling in the same direction are always in phase or always out of phase. However, if we look at the figure above, this is not so for waves of different frequencies: after about five oscillations, we see that the waves are definitely out of phase: when one is a maximum the other is a minimum. So, at this point, there will be destructive interference. But this condition will not last, either. As the waves keep going, they come back into constructive interference. Of course, we can always just add the two waves together to see what it will look like, as that is the essence of interference, Fig. 4.21.

Figure 4.21 Sum of waves in Fig. 4.20 showing a beat pattern.

If we listened to this wave, we would first hear the sound normally as it started with constructive interference. The sound would then fade away as it went through destructive interference, and then come back again. This fading in and out will occur repetitively and so will have a period of its own, separate from the two waves. In the example above, we can see that the period for the wave to go from loud to soft and back to loud is 1 s. So, the sound will seem to pulse with a frequency of 1 Hz. The effect is known as *beats*. In this example, the beat frequency is 1 Hz, which happens to be the difference in frequency between the two original waves. This is true in general: two waves with frequencies f_1 and f_2, when added together, will pulse or beat with a frequency equal to the difference between the two original frequencies, f_1 and f_2. Since this frequency should always be a positive number, we can write:

$$f_{beat} = |f_1 - f_2| \qquad\qquad (4.7)$$

For example, if we play two notes, one at 500 Hz and one at 502 Hz, they will beat together with a frequency of 2 Hz. Or, if we play a note at 500 Hz and another at 625 Hz there will be a beat frequency of 125 Hz. The phenomenon of beats is explored in Lab 4.3.

In terms of basic physics, this is really all there is to the phenomenon of beats. It simply means that when more than one frequency is present, one must consider, in addition to the main frequencies, the beat frequencies between all of the main frequencies.

Worked Example

A flute and a violin both tune their instruments to $A = 440$ indoors at a temperature of 20°C. They then go outside, where the temperature is 24°C. The violin's pitch does not change but the flute's pitch will, due to the change in temperature. What will be the beat frequency between the two instruments when they go outside?

Solution

Inside, according to Eq. 2.8, the speed of sound is $v_{in} = (331.3 + 0.6 \times 20)$ m/s = 343.3 m/s. Outside, the temperature will be $v_{out} = (331.3 + 0.6 \times 24)$ m/s = 345.7 m/s. So, according to Eq. 2.1, the frequency inside is $f_{in} = v_{in}/\lambda$, outside, it will be $f_{out} = v_{out}/\lambda$. The wavelength will not change, because that is set by the length of the flute. So, $f_{out} = f_{in}(v_{out}/v_{in}) = 440$ Hz(1.01) = 444 Hz. Therefore, the beat frequency will be (444 − 440) Hz = 4 Hz.

Human Perception of Beats

While the physics of beats is rather simple, how we respond to beats is probably the single most important consideration in how we perceive music. We will begin to unravel this rather strong statement here and return to it in much greater detail in Chapter 6.

It is not hard to see that beats are a complex phenomenon when it comes to perception. In the example just above, a note 500 Hz and one at 502 Hz will beat at 2 Hz. Again, this means the sound will get louder and softer with a frequency of 2 Hz. Now, 2 Hz, or 2 oscillations per second is a frequency one can keep up with. If a clock is ticking at 2 Hz, you can count the individual ticks.

In contrast to this, the notes at 500 and 625 Hz will beat at 125 Hz. Again, the sound will get louder and softer with a frequency of 125 Hz. But, how will the brain perceive this? Actually, this frequency is far too fast for the brain to follow individual oscillations, and so the brain does not interpret these beats as a periodic change in the loudness of the sound. In fact, under the right conditions, the brain interprets the beat frequency as its own frequency. In other words, you hear a new tone at 125 Hz. This is called a *difference tone*. We will return to this in Chapter 6.

Between these two extremes, there is a range of responses to beat frequencies:

1. Beat frequency is less than about 10 Hz.

   If the beat frequency is less than about 10 Hz, the two waves are very close in frequency and you will hear only one pitch (which is actually the average of the two frequencies). However, the loudness will vary with a frequency corresponding to the beat frequency and you will hear individual beats.

2. Beat frequency roughly in the range of 10–60 Hz.

   This range of frequencies is quite significant and problematic for the human nervous system. It is too fast for the individual events to be discerned, but too slow for the brain to ignore. Phenomena in this frequency range can be quite annoying. For example, if a film is played at 10 frames/s, you can identify the individual frames and it looks like a very fast slide show. If the film has more than 50 frames/s, you don't see the individual frames and the motion looks continuous. However, films at 20–30 frames/s look very jerky and annoying. The brain does not know whether to focus on individual frames, or try to interpret them as continuous motion. Similarly, strobe lights at 20–30 Hz can easily produce headaches. When it comes to sound, beats frequencies in this range sound rough and dissonant—they are simply unpleasant to listen to. The pitches are too far apart to consider as one pitch, but too close together to distinguish. The brain cannot follow the individual beats but it is not fast enough to interpret as a new pitch. *Simply put, when beats are present in this frequency range, the brain registers dissonance.*

3. Beat frequency is greater than about 60 Hz.

    When the beat frequency is high enough the brain can distinguish the pitches of the two notes and the beat frequency is either interpreted as a difference frequency, or simply two separate tones are heard.

These ranges depend somewhat on the absolute frequency of the two original notes, but, again, we will leave the details of this for Chapter 6.

As noted above, this range of frequencies is quite interesting for human perception and motor skills. If you hear taps in the range of 0–10 Hz, you can hear the individual taps. If the taps are greater than 60 Hz, they produce a pitch. However, in the range of 10–60 Hz, it is hard to interpret. However, it is also very hard to tap faster than about 10 Hz. The same is true of trills and vibrato—the periodic wavering of the pitch that can add warmth and richness to a note. So, motor skills are limited to this same frequency limit of about 10 Hz, but it is also right where interesting music effects are the most effective, such as trilling and vibrato. They occur just on the edge of what can be followed directly. Finally, it is interesting to note that the lowest note on the piano is at 26.5 Hz, which is also in this range of problematic frequencies. However, the lowest note on the piano does not sound so clean. It is often described as "muddy" and is not often used. In fact, the note itself can almost be described as dissonant on its own.

Consonance and Dissonance of Pure Tones and Complex Tones

To begin to understand how beats affect our perception of consonance and dissonance, let's start with a pitch of A = 440 Hz and examine what happens when we play a second note. For example, if the second note is at 450 Hz, you will hear one pitch at the average frequency of 445 Hz, but you will also hear beats at 10 Hz. This will not sound dissonant. However, if the second note has a frequency of 480 Hz, the sound will seem harsh and unpleasant and this is what is referred to as dissonance. There is a beat frequency at 40 Hz, but you cannot hear that distinctly. It is also hard to tell if there are two notes being played.

If the second note is around 510 Hz, the beat frequency will be 70 Hz. At this point, you will still call the notes dissonant, but not as bad as in the previous case. The two notes will now sound distinct and you can hear their individual pitches. Finally, for any frequency higher than about 530 Hz, the notes will be clearly distinct and consonant.

Now, how do these values compare to the intervals that we know? Well, a note a just major third above 440 Hz has a frequency of 550 Hz. The analysis above says that this should sound consonant, which it does. So far, so good. However, this analysis would say that *any* pitch above a major third would sound consonant. For example, a seventh should sound consonant, although this is definitely a dissonant interval. Even the tritone, which is a half step less than a perfect fifth should sound consonant by this reasoning, although it is the most dissonant interval of all. So something must be missing from this analysis.

Back in Section 3.2, we discussed pure tones and complex tones. A pure tone consisted of a single frequency equal to the fundamental of the note. Musically, you almost never hear a pure tone, except for a tuning fork. Indeed, that is the purpose of a tuning fork. The problem is that you have no control over the quality of a pure tone, so it is musically not very interesting. Thus, all instruments produce complex tones: in addition to the fundamental frequency, instruments also produce overtones, and the number and loudness of the overtones determine the tone quality. Of course, for melodic instruments, the overtones are all harmonics and in this case, the pitch of the note is not changed by the presence of overtones.

But what is the connection with consonance and dissonance? The analysis we discussed applies only to pure tones. When we considered a note at 440 Hz, we assumed that that was the only frequency present. Or when we played 440 and 550 Hz together, we assumed that there was only one beat frequency. In fact, the analysis *is* correct for pure tones. Two pure tones forming a tritone actually *do not* sound dissonant. So, while our analysis was correct, it just does not apply to musical tones, because musical tones never consist of pure tones.

So, we must go back and rethink what happens when we play two *complex* tones together. When we say that we play a complex note with a pitch of 440 Hz, we must remember that we also have frequencies of 880 Hz, 1320 Hz, 1760 Hz, ... present. Now, let's add a second complex note at 600 Hz. Before, we would have said that this sounds consonant, since the beat frequency with 440 Hz is 160 Hz. But, just as with the first note, the second note contains frequencies at 1200 Hz, 1800 Hz, ... Now, we must consider beat frequencies between *all* pairs of notes. So, we have 880 – 600 = 280 Hz, 1200 – 880 = 320 Hz, 1320 – 1200 = 120 Hz, and 1800 – 1760 = 40 Hz. Most of these beat frequencies are not a problem—if the beat frequency is relatively high, the brain is not concerned about it. However, the beat frequency at 40 Hz does cause a problem, as this is right in the range of the most dissonant beat frequencies. Thus, notes can sound dissonant not just because of the beat frequency between the fundamental frequencies of the notes, *but because of beats between any pairs of overtones.*

Because of the large number of overtones present in any complex tone and the greater number of pairs of overtones that must be considered, you might think that there are always going to be some dissonant beat frequencies. Basically, this is correct. Any two random complex notes will probably sound dissonant. But this is exactly why only special notes are used in a musical scale. Under certain conditions, the overtones of the pair of notes line up in such a way that there are no unpleasant beats.

For example, again consider a pitch of 440 Hz and now let's use the note a perfect fifth higher at 660 Hz. First, we need the harmonics of each note:

Pitch (Hz)	2nd harmonic	3rd	4th	5th
440	880	1320	1760	2200
660	1320	1980	2640	3300

In this case, the harmonics are either widely spaced and have large beat frequencies (e.g., 1760 and 1980 Hz), or the harmonics are exactly the same (e.g., both have 1320 as a harmonic). Generally, overtones which are close in frequency produce dissonance, *unless*, they happen to be exactly the same. This is why two notes exactly a perfect fifth apart sound consonant, while if one note is slightly out of tune, the pair sounds very dissonant. For example, if you play the note at 660 Hz just 10 Hz too high, at 670 Hz, the second harmonic will be at 1340 Hz. This will beat with the harmonic at 1320 Hz in the first note with a frequency of 20 Hz, and this will sound very unpleasant. So, consonance and dissonance are really determined by the frequencies of the overtones of the two notes. Again, we will discuss this more in Chapter 6, including various experiments in Labs 6.1 and 6.2.

4.6   Light Waves

Throughout this course, we have mainly focused on sound waves, as that is the most important wave for music. We also have considered waves on strings and on metal bars, as each of these gives rise to a whole class of musical instruments. However, there is another type of wave that is extremely important to physics and science and our everyday life, and that is light waves. Given what we already know about waves, in general, adding light to our list of waves will not be so difficult, although light does have some rather unusual properties. As discussed in Section 4.1, all waves share a variety of properties, and we will examine light waves along these same lines.

Waves Require a Medium

The medium for most waves is usually apparent. For a string wave, the medium is a string. For sound waves, the medium is air. What is the medium for light? We know that we can detect the light from distant stars and galaxies, so apparently light can travel through empty space. So, does this mean that light does not require a medium, or does it mean that empty space is not so empty, after all? As soon as it was realized that light is a wave, the question of its medium has plagued physicists up to this day. At this point, the answer seems rather clear, as light is now quite well understood, but the conclusion is startling: light, indeed, does require a medium and, thus, space is not empty. Space is filled by a "sea" of "virtual" positrons and electrons that, normally, we are not aware of. What is really meant by the preceding sentence goes far beyond the scope of this book. However, the properties of this unusual medium are still an active area of research.

Even if the medium is still not well understood, the restoring force is. The restoring force for light is comprised of the more familiar electric and magnetic forces. The electric force drives electricity through wires and the magnetic force is what makes magnets attract or repel each other.

Waves Carry Energy with a Certain Velocity

As we have discussed before, waves really have an existence of their own because they can carry energy. Moreover, waves are different from normal objects because they are always in motion—they are always moving with the speed characteristic of that particular wave. Both of these features make light waves very useful in a wide range of situations.

At the most basic level, nearly all of the energy that keeps the Earth warm and keeps us alive comes, directly or indirectly from the Sun, in the form of light. Indeed, the amount of energy striking the Earth in the form of light is staggering: 5×10^{17} joules of energy *per second*. Even if only a tiny fraction of this can be harnessed, all of our energy problems would be solved.

The light also delivers energy very quickly: the speed of light is 3×10^8 m/s, or, perhaps the more familiar 186,000 miles/s. Another odd property of light is that this speed sets the absolute speed limit for everything in this Universe—nothing can travel faster than light. This speed is unimaginably fast, but it is not instantaneous. In fact, the Sun is so far away, that is takes 8 min for the light of the Sun to reach the Earth. Because of the relationship $f\lambda = v$, the very large speed of light affects the range of wavelengths and frequencies that we encounter with light.

Light waves that we come across in daily life cover a huge range of frequencies. Starting at the "low" frequency end of the spectrum of light, we find radio waves. When you set your AM radio to 1040, you are really selecting the frequency of radio waves that you want to receive. Each radio station is given its own frequency to keep its transmissions separate from everyone else's. The 1040 corresponds to 1040 kHz, or about 1 MHz. If you listen to 104 on the FM dial, this corresponds to 104 MHz.

Although these frequencies are very large compared to the frequencies we are used to from sound waves, they do represent the low end of the spectrum. As we get up to frequencies of around 1 GHz (10^9 Hz) we find the waves responsible for TV transmission and cell phone communications.

Even higher frequencies lead to microwaves, as in the waves used in microwave ovens. Here, it is quite clear that light waves are being used for their ability to carry energy.

All the uses of light waves are too vast a topic to cover here, but as the frequency increases, we encounter infrared, visible, ultraviolet, and x-ray waves. Although the uses and applications of these waves are incredibly diverse, they are all as similar to each other as the different notes on the piano are. All of these waves are simply light waves of different frequencies.

Transverse or Longitudinal?

Waves are classified into two groups: transverse or longitudinal. This indicates how the direction of motion of the medium is related to the direction of motion of the wave, itself. Remember, for a string wave, the string moves perpendicular to the string, while the wave travels along the string. Thus, a string is a transverse wave. For sound waves, the air molecules move along the motion of the wave, and hence, sound is a longitudinal wave.

What is light: longitudinal or transverse? This is obviously a hard question to answer, since we hardly understand the medium that it is moving in. However, we can deduce what it must be. For a longitudinal wave, there is only one direction for the medium to move along—the direction that the wave in moving. However, for transverse waves, there are always two directions that are perpendicular the direction of the wave. For a string wave, this may be up and down and back and forth, if the wave is traveling left to right. So, if we can determine whether there is only one direction for the medium to vibrate, or two, we can determine if the wave is longitudinal or transverse.

As it turns out, there are two directions for light waves to oscillate and, consequently light is a transverse wave. Although this may all seem rather obscure, the item needed to prove this is quite ordinary—a pair of polarized sunglasses will work. When you put on a pair of sunglasses the amount of light reaching your eyes is cut approximately in half. This makes things more comfortable for your eyes, but how is the light level reduced? It may be that the lenses simply absorb half the light, and, in fact, regular non-polarized sunglasses do just that. If you place a second pair of regular sunglasses right in front of the first, the light level will again be cut in half, and now only a quarter of the light will reach your eyes.

However, if you perform the same experiment with the polarized sunglasses, something completely different happens. If you look through one pair of polarized glasses, the light is cut in half, as we have said. But, now, if you look through a second pair, no additional light is lost. One pair of polarized glasses cuts just as much light as two. Now for the strange part: if you rotate the second pair of polarized glasses by 90° with respect to the first pair, you see nothing. All of the light is gone. As bizarre as this seems, it has a perfectly reasonable explanation related to the fact that light is a transverse wave. Lab 4.4 explores the polarization of light.

As we discussed, transverse waves have two directions that they can vibrate in that are perpendicular to the direction that the wave is traveling. These two directions are called polarizations (which is where the term for polarized sunglasses comes from). If you are looking out at a distance object, the light waves that you see are traveling straight at you. The two polarizations, or directions that the light vibrates in, are then up-down (or vertical polarization) and left-right (or horizontal polarization). Natural light contains roughly equal amounts of each polarization. What polarized glasses are able to do is let one polarization through and block the other. For example, polarized sunglasses are set to absorb the horizontal polarization, which is called a vertical polarizer. So, half the light is blocked and the scene is less bright. However, if you look through a second pair of polarized glasses, as well, the horizontal polarization has already been removed, so there is no further change to the light. If, on the other hand, you rotate the second pair of glasses by 90°, then the second pair will remove the vertical polarization. In this case, both the horizontal and vertical polarizations are gone, and there is no light left.

Of course, no one goes around wearing two pairs of sunglasses, so why are polarized sunglasses useful? This is easily demonstrated in Fig. 4.22 with a bowl of water and a bright lamp.

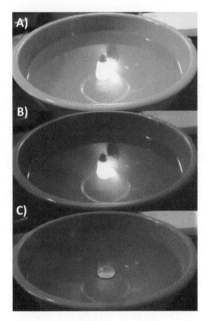

Figure 4.22 (A) Bowl with no polarizer. (B) Bowl with horizontal polarizer. (C) Bowl with vertical polarizer.

The first image is taken with no polarizer and there is a strong reflection of the lamp off of the water. The second image is taken with a horizontal polarizer (i.e., blocks vertical polarization). Overall, the image is somewhat darker, as roughly half of the light is removed by the polarizer, but the strong glare is still there. It turns out that the glare from reflected light is strongly polarized in the horizontal direction, thus the horizontal polarizer does not affect it. Finally, the last image is taken with a vertical polarizer. In this case, the glare is almost completely removed, revealing an object at the bottom of the bowl. Polarization filters like this are often used in photography.

Light Waves Show Interference

One of the simplest ways to show that something is a wave is to show that it produces interference. Recall that in Section 4.4 we discussed that fact that electrons must be waves because they show constructive and destructive interference. In the same way, light must also be a wave because it shows interference.

Interference in Reflection

One of the more useful applications of the interference of light in reflection is to produce antireflection coatings for glasses. Reflections off of glass can be suppressed by placing a very thin transparent coating on top of an eyeglass lens, for example, Fig. 4.23.

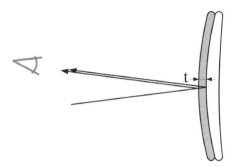

Figure 4.23 Interference of light in reflection. A thin transparent layer (in grey) with thickness t is placed on a lens.

Normally, a piece of glass, like an eyeglass lens reflects a fair amount of light, about 4%. Although this may seem small, it can be distracting when talking to someone wearing glasses, especially if there are bright lights around. The reflection is particularly noticeable in photographs or on television because of the bright flash or camera lights. These reflections would cause problems within a camera, as a camera lens is not a simple lens like in eyeglasses. Rather, a good camera lens is actually composed of up to ten individual lenses. If each lens reflected some light, there would be a lot of light bouncing around inside the camera, washing out the photograph. Thus, scientists came up with a way to suppress the reflections using the principle of interference.

Looking at Fig. 4.23, a single wave comes along and strikes the eyeglass lens. A small part of the wave reflects off of the first surface—the front of the coating. The rest of the wave is transmitted. However, a small part of the transmitted wave then reflects off of the lens, itself. Thus, two waves are now traveling back and can interfere. The goal is to make them interfere destructively.

Following the procedure above we find that $2t = \lambda/2$ for destructive interference. This means that the thickness $t = \lambda/4$.

Technically, t refers to the optical path of the coating, because of a property of the glass called index of refraction, but these details are not import for understanding the interference effects.

This is our final result: a simple lens will reflect about 4% of the light striking it. However, if you place a transparent coating over the lens, the reflected light can be reduced almost to zero through destructive interference. In the range of light that we can see, the wavelength is roughly 0.5 μm. So, the thickness of the "antireflection coating" needs to be about a quarter of this, or about 0.125 μm. Of course, this is very thin, but sophisticated techniques have been developed to apply coatings of this thickness very precisely. It does add about $50–100 to the price of the lens, but it can be worth it. As stated above, in more complex optical systems, like camera lenses, zoom lenses, binoculars, telescopes, etc., antireflection coatings are absolutely essential. Here is a photograph of two pairs of glasses, Fig. 4.24. The upper one has the antireflection coating, while the lower one does not.

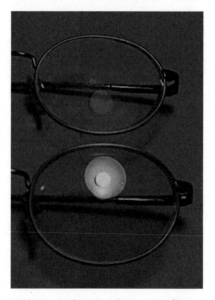

Figure 4.24 Glasses with and without antireflection coatings.

One major problem with these coatings is that they must be kept very clean. If you leave a fingerprint on a lens, the thickness of the oil layer that is left behind can easily be in the range of μm's. Notice that if the coating thickness is $t = \lambda/2$, we will have *constructive* interference, which will make the reflected light *brighter*. This is why the thickness must be controlled so carefully, and why dirt on the surface can so easily disrupt the conditions for destructive interference.

Creative entrepreneurs have found a way to take advantage of this delicate balance. Imagine that you apply a coating to a surface that is not flat, but has been patterned in some way, like in Fig. 4.25.

Figure 4.25 Cross section of wrapping paper which produces colorful patterns through interference.

The spots where the thickness produces destructive interference will appear dark, while the spots where there is constructive interference will appear bright. This produces interesting patterns of light and dark. In fact, it is a bit more complicated since the conditions for constructive and destructive interference depend on the wavelength of the light. The wavelength of light determines the color that you perceive. So, different colors will show different patterns of constructive and destructive interference giving a striking "rainbow" effect, as in Fig. 4.26. This effect has been used to make shiny colorful wrapping paper and it is also how data are stored on compact discs (CDs).

Figure 4.26 Colors produced through interference of light.

Interference of light in reflection produces the colors that you often see in puddles in the road. Just like glass, water reflects about 4% of the light striking it. Now, roads contain a lot of oil and some of this oil gets into the puddle. Since oil is lighter than water it rises to the surface and spreads out over the surface of the water, producing a thin film. This film of oil acts like the coating that we discussed above. However, the thickness is not precisely controlled, so it will not produce purely constructive or destructive interference. However, it does affect how much light of each color is reflected. This produces the patterns of very pure colors, like a rainbow, that you often see when looking at a puddle of water.

Another situation to demonstrate the interference of light in reflection is similar to sound bouncing off steps (see Fig. 1.1). A "set of stairs" for light is called a diffraction grating and consists of fine grooves on a piece of glass, where each groove reflects some light. In this case, it is possible to vary the length of the "steps" simply by holding the grating at an angle θ, as in Fig. 4.27.

In this case, the length of the steps is $D = G\cos(\theta)$, where G is the spacing of the grooves. For constructive interference, we have $\lambda/2 = D = G\cos(\theta)$, or $\lambda = 2G\cos(\theta)$. This provides a very useful way of measuring the wavelength of light, simply by tilting the grating until the light reflects straight back, due to constructive interference. This is an example of a more general phenomenon called diffraction, which can be demonstrated in Lab. 4.5. Diffraction describes how light would reflect off a grating at any angle.

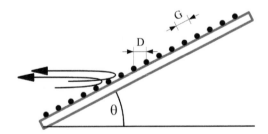

Figure 4.27 Grating for light with a line spacing of G.

Interference Patterns

There are many ways to show the interference of light and many applications based on interference, so we can only list a few, here. One of the most famous demonstrations of interference uses a "double slit". The double slit is produced by creating two very fine slits in an opaque mask, as shown in Fig. 4.28.

Figure 4.28 Mask with two fine slits to show the interference of light waves.

If you shine a bright light like a laser beam through this pair of slits, you produce two sources of light, very much like the two speakers in Section 4.4. From the top, it would look something like Fig. 4.29.

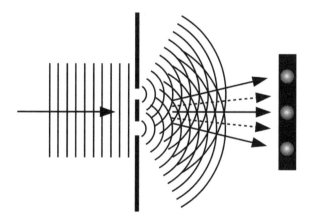

Figure 4.29 Set-up to observe the interference of light waves.

Figure 4.30 shows two-slit interference patterns using the same slits but different wavelengths or colors of light. Looking at the number of spots between the vertical lines, we see that the spacing of the points of constructive interference gets smaller as the wavelength gets small. This is in agreement with Eq. 4.6.

The two waves will interfere, just like the sound waves. In fact, the spacing of the points of constructive or destructive interference will follow exactly the same formula as in Section 4.4. Of course, you must use the correct wavelength for light waves. Also, the interference pattern tends to be a series of bright spots, as laser beams are generally quite small in diameter (Fig. 4.30).

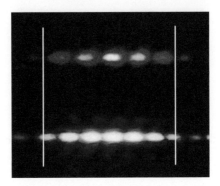

Figure 4.30 Two-slit interference of light with two wavelengths: 650 nm (red) and 532 nm (green).

Worked Example

Equation 4.6 qualitatively predicts that as the wavelength of light gets smaller, the spacing of point of constructive interference gets smaller, as seen in Fig. 4.30. The question here is does 4.6 *quantitatively* predict how the spacing changes with wavelength.

Solution

For the red laser in Fig. 4.30, there are 5 spots within the white lines. We don't know the physical distance, we can say that the distance between the lines represents one unit, so the spacing, $y_{red} = 1/5 = 0.2$. Going back to Chapter 1, we can write: $y_{red} = \frac{\lambda_{red} L}{D}$. Similarly, $y_{green} = \frac{\lambda_{green} L}{D}$. Dividing these two equations gives $\frac{y_{red}}{y_{green}} = \frac{\lambda_{red}}{\lambda_{green}}$, or $y_{green} = \frac{\lambda_{green}}{\lambda_{red}} y_{red} = \left(\frac{532}{650}\right) 0.2 = 0.16$. A spacing of 0.16 predicts $1/0.16 \sim 6$ spots within the white lines for the green light which, indeed, quantitatively agrees with Eq. 4.6.

Interference in Time

Finally, interference of light waves in time has a very important application and that is radio transmission. Say you want to broadcast a note at 440 Hz. If you just play this note into a speaker sitting on top of a building, you would have to play it very loudly for others to hear it. Instead, you decide to use radio waves to transmit this sound. First, you must choose a radio frequency, like 900 kHz. This identifies you as the source of the sound, and anyone who wants to listen to you tunes into this frequency, but, so far, they will not hear anything. To broadcast the note at 440 Hz, you actually transmit a second wave at 900 kHz + 440 Hz, or 900,440 Hz. Now, when someone tunes to 900 kHz, they will detect a beat frequency of 440 Hz. Both waves can be sent directly to a speaker. The speaker does not respond to the high radio frequencies, but it does respond to the acoustic beat frequency at 440 Hz. As a result, you will hear the 440 Hz tone. Of course, you can add as many frequencies as you want to create a full musical sound. This is called AM, or amplitude modulation, and is the method used by AM radio stations. There is a rather different technique, called frequency modulation, or FM, which we will not discuss, here.

4.7 Problems

Section 4.1

4.1 Briefly discuss three properties of waves that make them different from objects.

4.2 Which of the following statements are true?

A. A restoring force is associated with the medium, not the wave.
B. The louder a wave the faster it travels.
C. Only objects carry energy, waves do not.
D. Waves bounce off each other, like objects.
E. A wave can be in multiple places at once.
F. Waves require a medium to exist.
G. Waves can be considered disturbances in a medium.

4.3 If you increase the tension of a string, the wave velocity will:

A. Depends on the loudness.
B. Decrease.
C. Stay the same.
D. Increase.

4.4 Indicate if the following waves are A. transverse; B. longitudinal; C. both; or D. neither.

(a) Surface waves on water.
(b) Sound waves.
(c) Waves on strings.
(d) Waves on a slinky.

4.5 What is the restoring force for each of these media, and does it support longitudinal waves, transverse waves, or both? Air, Water, Slinky, Drum.

Section 4.2

4.6 If I am standing in the middle of a room and there is one speaker a distance R_1 away from me and another speaker R_2 away from me, what does $|R_1 - R_2|$ depend on?

A. The frequency of the sound.
B. The locations of the speakers.
C. Whether I have constructive or destructive interference.
D. All of the above.
E. None of the above.

4.7 If I am standing in the middle of a room and there is one speaker a distance R_1 away from me and another speaker R_2 away from me, what determines whether I have constructive or destructive interference?

A. The wavelength of the sound.
B. $|R_1 - R_2|$.
C. How far apart the speakers are from each other.
D. (A) and (B)
E. (A), (B), and (C)

4.8 Imagine that you are standing in front of two speakers as shown below.

If I play a frequency of 300 Hz and the speakers are separated by 57 cm (i.e., $D = 57$ cm), will I have constructive or destructive interference or something in between? Use 343 m/s for the speed of sound.

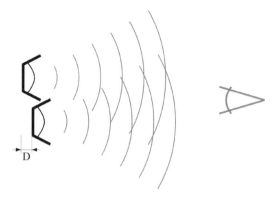

A. Constructive.
B. Destructive.
C. In between.
D. Cannot determine from the information provided.

4.9 Imagine that you are standing in front of two speakers as in Problem 4.8.

(a) If the wavelength of the sound is λ = 10 cm and the second speaker is moved back by a distance of 20 cm, do we have constructive or destructive interference?

(b) How much further back must we move the second speaker to reverse the situation in a)?

(c) If we play a tone with a frequency of 550 Hz, what separation between the speakers will produce destructive interference? (There are many answers, just give one).

4.10 (a) We start with two speaker emitting pure tones at 1500 Hz. Your distance from each speaker is R_1 and R_2, respectively:

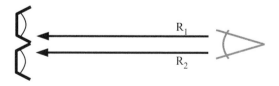

When the speakers are very close to each other, do we have constructive or destructive interference?

(b) We now move speaker 2 backwards a distance D:

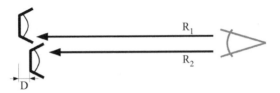

How far must we go to switch from constructive to destructive or destructive to constructive interference?

(c) Now we move speaker 2 to the side by a distance D:

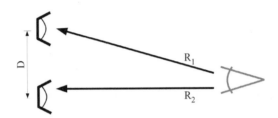

Again, how far must we go to switch from constructive to destructive or destructive to constructive interference, if R_2 = 2 m? Calculate this exactly. (Hint: you know R_2. Figure out what R_1 should be to give the correct interference condition. Then use the Pythagorean Theorem to get D.)

4.11 Consider the arrangement of speakers shown in Fig. 4.10, which are facing each other.

(a) If we use a frequency of 3500 Hz, what is the wavelength of the sound wave?

(b) If we are exactly half way between the speakers, what is $R_2 - R_1$? Will we have constructive or destructive interference?

(c) How much must we move to the left to be in the opposite situation as in part b)? (Both R_1 and R_2 are going to change.)

(d) How does the answer to c) relate to what you know about the distance between nodes and antinodes on a standing wave?

Section 4.3

4.12 In Fig. 4.12, assume the gap in the wall is 9 cm.

(a) Which of the following frequencies will show constructive interference? Mark all that apply. Use 343 m/s for the speed of sound.

(A) 317 Hz; (B) 635 Hz; (C) 952 Hz; (D) 1906 Hz;
(E) 2858 Hz; (F) 3811 Hz; (G) 4764 Hz; (H) 5717 Hz

(b) Which of the following frequencies will show destructive interference?

(A) 317 Hz; (B) 635 Hz; (C) 952 Hz; (D) 1906 Hz;
(E) 2858 Hz; (F) 3811 Hz; (G) 4764 Hz; (H) 5717 Hz

4.13 (a) A wall consists of a strong exterior wall and a light interior wall with a gap in between for insulation, as in Fig. 4.12. If the gap between the interior and exterior wall is 10 cm, what frequency of sound will be strongly reflected by the whole structure?

(b) What frequency of sound will exhibit destructive interference and get absorbed by the wall?

4.14 Suppose you are in a Greek theater surrounded by steps with a depth $D = 3$ feet, as in Fig. 1.1.

If you clap your hands loudly, you will get an echo from the steps. However, the echo from the lowest step will arrive first, then the echo from the next step and so on. All of the echoes will sound like a pitch. What pitch will this be?

4.15 You are singing in front of some steps with a width $D = 0.746$ m (see Fig. 1.1).

(a) If you produce a frequency of 230 Hz, will that show constructive or destructive interference when it reflects off of the steps?

(b) Now you sing at 345 Hz. Will this frequency show constructive or destructive interference?

4.16 (a) Assume that you are standing in front of some steps and that when you sing a note with a frequency of 450 Hz the echo sounds loud due to constructive interference (see Fig. 1.1). What is the length of each step?

(b) What other frequencies will also sound loud due to constructive interference?

(c) What kind of series do these frequencies form?

4.17 (a) Assume that you are standing in front of some steps with a width of 2 m (see Fig. 1.1). What frequencies will sound loud due to constructive interference?

(b) What kind of series do these frequencies form?

(c) What frequencies are not heard due to destructive interference?

(d) What series do these frequencies form?

4.18 You are playing a low "A" at 110 Hz in front of a (solid) wall that is 0.78 m away from you.

(a) Will the reflected sound constructively or destructively interfere with the direct sound you are producing?

(b) Will the second harmonic constructively or destructively interfere?

4.19 You are sitting in the audience listening to a tuba play a low $E = 137.5$ Hz. Some of the sound comes to you directly, while some reflects off of the wall behind the orchestra.

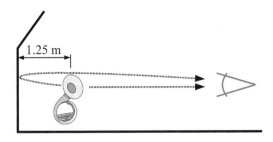

(a) Will the direct sound and reflected sound interfere constructively or destructively where you are sitting?

(b) If the tuba plays a note a perfect fifth higher, will there be constructive or destructive interference?

4.20 (a) Assume that while singing in front of some steps you hear no echo at a frequency of 500 Hz because of destructive interference. What is the length of each step?

(b) What frequencies would you hear loudly?

Section 4.4

4.21 If you want to make a pattern of interference fringes more finely spaced (i.e., make spacing of the fringes smaller), you should:

    A. Decrease the wavelength.

    B. Decrease the separation of the sources.

    C. Decrease the distance to the point of observation.

    D. (A) and (B).

    E. (A) and (C).

4.22 In the following arrangement, what frequency will produce a spacing of 1 m between points of constructive interference? (Use 343 m/s for the speed of sound.)

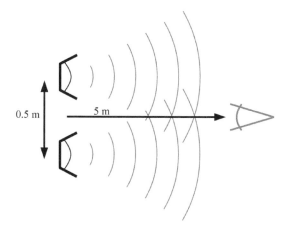

    A. Any frequency.

    B. 440.0 Hz.

    C. 1380.1 Hz.

    D. 3430.0 Hz.

    F. (B) and (C).

4.23 Two speakers project sound waves with a wavelength of 5 cm in all directions and are placed 5 cm apart. If you walk in a circle around the speakers, mark the points with an *x* where the sound would be the loudest. Then mark with a • where the sound would be the softest.

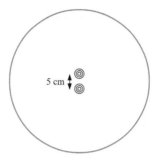

4.24 (a) Consider sound waves with a frequency of 3430 Hz. What is the wavelength of this sound wave?

(b) In each case below, two speakers are placed a certain distance apart, as labeled. Mark on the circle approximately the locations where you would find constructive interference (X) and destructive interference (•).

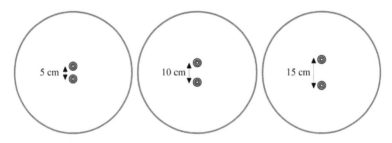

4.25 Consider the following interference patterns:

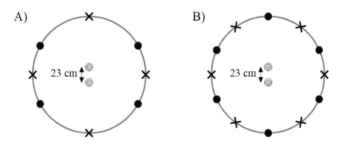

(a) In diagram (A), what are the wavelength and frequency of the wave?

(b) In diagram (B), what are the wavelength and frequency of the wave?

4.26 Ultrasound has frequencies around 40,000 Hz. If you have a frequency of 41188 Hz, what wavelength does this correspond to, in centimeters? (Use 343 m/s for the speed of sound.)

4.27 (a) Ultrasound has frequencies around 40,000 Hz. What wavelength does this correspond to?

(b) If you stand 3 m away from two speakers separated by 20 cm, how much would you have to move your head to go from dead spot to the next? That is, what is the distance between points of destructive interference?

Section 4.5

4.28 A note has a pitch of 356 Hz. What is the beat frequency between this note and a note a whole step (9/8) higher?

4.29 In Chapter 3, we saw that the harmonics of a string in a grand piano may not be perfect. Consider "C" = 256 Hz.

(a) What is the frequency of a "C" three octaves higher?

(b) Assume the 7th overtone of this string is at 8.05f. What is the frequency of the 7th harmonic?

(c) What is the beat frequency between the note found in (a) and the 7th harmonic found in (b)?

4.30 (a) Take a note with a fundamental frequency of 420 Hz. What note is an interval of a fourth higher?

(b) What are the frequencies of the overtones that these notes have in common? (Assume that the overtone series of each note is a complete harmonic series.)

(c) If the second note is out of tune by being 3 Hz too high, what is the beat frequency between the first common overtones?

4.31 (a) Consider a note with a frequency of 1000 Hz. What note is an interval of a fifth higher?

(b) What is the beat frequency between these two notes?

(c) This beat frequency is rather high and the ear will not think of it as a "beat". Rather, the ear will simply interpret it as a new pitch. What interval below the original note will the beat frequency be at?

4.32 (a) Two clarinetists play the same note with a frequency of 440 Hz at $T = 10°C$. What is the wavelength of the tone?

(b) Now, one clarinet warms up to $T = 12°C$. If the wavelength stays the same, what is the pitch now?

(c) What will be the frequency of the beats between the two clarinets?

Remember, the speed of sound is given by Eq. 2.8.

4.33 (a) The note 'C' on a piano has a frequency of 264 Hz. If we go up by an interval of a fifth and down by an interval of a just third, you have a new interval called a minor third. What is this ratio? What frequency is a minor third above 'C'?

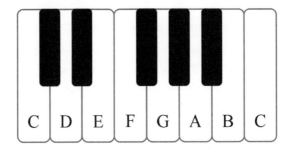

(b) Pythagoras would create a minor third by going up a fifth and coming down by a Pythagorean third. What is this ratio? What is the frequency of the note?

(c) What is the beat frequency between the notes found in parts (a) and (b)?

Section 4.6

4.34 Identify three similarities and three differences between light waves and sound waves.

4.35 Section 4.5 shows interference with light using a laser pointer and a mask with two very fine slits placed very close together. The formula for the spacing of the fringes (points of constructive interference) is $y = \lambda L/D$, where D is the separation of the slits, L is the distance to the point of observation and y is the spacing of the fringes. Which of the following has reasonable numbers for y, L, and D?

A. $L = 6$ m, $D = 0.2$ mm, $y = 2$ cm
B. $L = 20$ cm, $D = 1$ m, $y = 2$ cm
C. $L = 1$ mm, $D = 6$ m, $y = 0.1$ mm
D. $L = 50$ cm, $D = 0.1$ mm, $y = 0.5$ mm

4.36 (a) Your favorite radio station uses a frequency of 104.1 MHz. What wavelength does this correspond to?

(b) If you are standing near a wall, your radio will pick up the direct signal from the station, as well as a reflection from the wall. How close to the wall would you have to be to find destructive interference?

4.37 The eye can see over a range of wavelengths from about 400 to 700 nm (1 nm = 10^{-9} m). We want to design the antireflection coating for the wavelength right in the middle of this range.

(a) What is the middle wavelength, and how thick should the coating be to create *destructive* interference in the reflected waves at this wavelength?

(b) For the thickness determined in part (a), what wavelength will show *constructive* interference?

(c) The antireflection coating is actually effective from about $(4/3)\lambda$ to $(3/4)\lambda$, where λ is the design wavelength in part (a). Does this range cover the range that the eye can see?

4.38 Cavity modes of lasers are very similar to the acoustic cavities that we have been working with. However, there are some differences.

(a) Consider a one-dimensional cavity of light with a length of 50 cm. If the boundary conditions are the same as a closed-closed acoustic cavity, what is the frequency and wavelength of the fundamental mode?

(b) This fundamental mode has a much lower frequency than the light generated by the cavity. If the light has a wavelength of 500 nm, what is the frequency of the light? What mode will this correspond to? (The mode number will be very large).

4.39 The fundamental mode of the laser cavity in the previous question has a much lower frequency than the light generated by the cavity. In this problem, if the laser cavity has a length of 111 cm and the cavity produces light with a wavelength of 395 nm, what is the number of the mode that the laser is in? (The mode number will be very large). Use exponential format for the answer. The speed of light is 3×10^8 m/s. Note that the length is given in cm and the wavelength in nm.

4.40 (a) A lightning bolt hits the ground 1 mile away from you. How long does it take the flash of light to reach you?

(b) How long does it take the sound to reach you? What is a simple relationship to convert from seconds between the flash and the sound to figure out the distance?

(c) Microwave ovens use a frequency of 2.45 GHz. If this radiation formed a standing wave in your microwave oven, what would be the distance between nodes and antinodes? How does this distance compare to the size of a potato? This is why there are rotating platters in such ovens.

4.41 Light, which is a wave, travels through space with a speed of 3×10^8 m/sec. Find the frequency of the following: (1 nm = 10^{-9} m)

(a) Radio waves with $\lambda = 120$ m.
(b) Red light with $\lambda = 730$ nm.
(c) Ultraviolet light with $\lambda = 300$ nm.
(d) Microwaves with $\lambda = 3$ cm (used in police radar).

4.42 Two slits separated by a distance D create the two sources of light. See Fig. 4.29. If the wavelength of light is 600 nm (600×10^{-9} m), the slit separation is 0.1 cm and the distance L to the screen is 2 m, what is the separation of the points of constructive interference.

4.43 (a) The two light waves are formed by two slits in a mask with a separation of 0.2 mm (1 mm = 10^{-3} m). See Fig. 4.29. If we measure the interference pattern 2.3 m away from the mask and find that the spots of constructive interference are 4.5 mm apart, what is the wavelength of the light?

   (b) If the separation of the slits is made smaller by a factor of 1.5, what would be the separation of the spots of constructive interference?

4.44 While making soap bubbles you notice that with the light behind you, the reflected light from the soap bubble looks green. How thick is the soap film? (Green light is around 530 nm).

4.45 If we have a grating whose groove spacing G is 10^{-6} m at an angle of 25°, what wavelength of light will be reflected straight back along the horizontal direction? See Fig. 4.27.

4.46 (a) A grating has 500 steps/mm. What is the distance between the grooves, G? See Fig. 4.27.

   (b) You shine a laser with a wavelength of 630 nm at the grating. At what angle will the light reflect straight back at you?

   (c) At how many different angles will this happen?

4.47 (a) A grating has a line spacing G of 0.6 μm (1 μm = 10^{-6} m). See Fig. 4.27. If the grating is held at an angle of 30°, what is the longest wavelength of light that will be reflected straight back?

   (b) What other wavelengths will be reflected straight back?

4.48 Once again, we have a grating to reflect light. See Fig. 4.27. This time we want to know which angles of the grating will not reflect any light straight back.

   (a) In symbols, write down the path difference, $R_2 - R_1$, between two waves which are reflected off of two adjacent grooves.

   (b) In symbols, write down the general condition for having destructive interference. Remember, you can always add multiples of the wavelength.

(c) Now, set these expressions equal. If $G = 6 \times 10^{-7}$ m, and $\lambda = 530 \times 10^{-9}$ m, what angles will give destructive interference? (There are only two such angles).

4.49 Hydrogen produces light with characteristic frequencies. The first four wavelengths are 121.5 nm, 102.6 nm, 97.26 nm, and 94.98 nm (1 nm = 10^{-9} m).

(a) Convert these wavelengths to frequency.

(b) Write these frequencies as ratios with the fundamental (for example, a metal bar is 1.00f, 2.54f, 5.40f, etc.).

(c) Show that the ratios for the overtones are given by $(4/3)(n^2-1)/n^2$. You need to find the integer n for each ratio.

(d) Using the formula $(4/3)(n^2-1)/n^2$ find the interval that the first two overtones make with the fundamental.

(e) For very large n, what interval does this ratio approach?

(f) Predict the next wavelength in the above hydrogen series.

Chapter 5

Applications: Feedback and the Doppler Shift

5.1   Gain and Feedback

In this chapter, we will consider two phenomena related to waves that have numerous important technological and scientific applications: feedback and the Doppler effect. While you may be somewhat familiar with the Doppler effect, you may be wondering what I mean by feedback and why it is important to anyone. The answers to these questions will be the focus of the next two sections.

One example of feedback that most people are familiar with is the screeching noise that sometimes is created when someone is talking into a microphone and the volume is turned up to high. It can also occur when a musical group is playing and, again, the microphones are turned up too high. Actually, some groups use feedback intentionally to produce a particular kind of sound.

So, what exactly is feedback and how can we use it in a controlled way? Say we want to measure the resonant frequencies of an open-open tube. We can generate an audio signal that sweeps through a range of frequencies and send that to a speaker near the tube, as in Lab 2.3. A microphone detects the sound and an amplifier increases the signal to where a computer can record it. The signal will increase significantly when the input frequency matches the resonant cavity modes and, in this

The Music of Physics: An Introduction to the Harmonies of Nature
George N. Gibson
Copyright © 2024 Jenny Stanford Publishing Pte. Ltd.
ISBN 978-981-4968-99-7 (Hardcover), 978-1-003-48481-3 (eBook)
www.jennystanford.com

way, the resonances can be measured. Figure 5.1 shows this schematically.

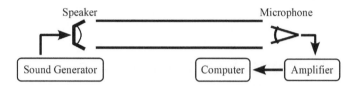

Figure 5.1 Open loop to measure cavity response.

It is important to note that this is a *linear* system, which means that the signal moves in a single direction and elements farther down the signal path do not influence earlier elements. For example, the output of the microphone does not affect the sound generator. In addition, if the loudness of the sound generator is doubled, the resulting output of the microphone will double.

In a system with feedback, the flow of the signal is not so simple: later elements can affect earlier elements. For example, what would happen if we took the output of the amplifier and connected it directly to the input of the speaker? Figure 5.2 shows the resulting flow chart.

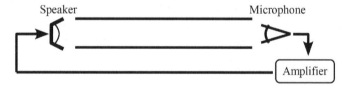

Figure 5.2 Closed loop to produce feedback on a cavity mode.

This arrangement has fewer elements, and so it may appear to be simpler than the first, but since it is not connected in a simple way, its behavior is actually much more complicated. However, we can get an idea of where the term "feedback" comes from. In this example, the output of the amplifier is "fed back" into the input of the speaker. This makes it difficult to figure out what this arrangement is going to do. The output of the amplifier box drives the speaker, which creates waves in the cavity, which produces a signal in the microphone, which serves as the input to the amplifier, which changes the output of the amplifier—and the whole cycle starts over again. Because the signal travels through the flow chart in a loop, these systems are often referred to as "feedback loops".

Now that we have some idea of what a feedback loop will do, let's start to analyze it. Since it is a loop, we can start anywhere and follow the signal around the loop. Let's just assume, in the example above, that there is a signal of 0.1 V on the speaker. This will produce a wave of a certain amplitude in the tube. This wave, in turn, will produce an output signal from the microphone. Much of the energy in the initial signal going into the speaker will be lost in the cavity and the microphone, so the signal coming out of the microphone will be much smaller. As an example, let's say that it will be 100 times smaller. So, instead of 0.1 V going into the speaker, we will only have 0.001 V coming out of the microphone. This reduction in the signal level is called *loss*. It comes about because microphones only produce very small signals, which we will need to amplify. So, let's say that the amplifier increases the signal by a factor of 10. This helps the signal out of the microphone a lot, increasing it from 0.001 to 0.01 V. Now, this signal is feed back into the speaker to complete the feedback loop.

But, what has happened? We started by saying that there was 0.1 V on the speaker, and after going through the loop, we only came up with 0.01 V, or a signal 10 times smaller. So, when this new signal goes through the loop, it will be 10 times smaller, again, or 0.001 V. The signal keeps going around the loop getting 10 times smaller each time. Very quickly, there will be nothing left to the signal. Since there is actually no signal to start with, there never will be a signal.

Now we will change the situation. Instead of increasing the signal out of the microphone by only a factor of 10, let's make the increase a factor of 200, which is more typical of a microphone amplifier. Starting again with a signal of 0.1 V on the speaker, there will again be a signal of 0.001 V out of the microphone. However, the amplifier will now increase this to 0.2 V and send it back to the speaker. This signal is larger than the one we started with. If we have 0.2 V on the speaker, there will be 0.002 V out of the microphone, which will be amplified to 0.4 V and sent back to the speaker. This is now a very different situation than we had before. Instead of the signal getting smaller on each loop, it gets larger. In fact, no matter how small the signal is originally, it will grow and grow until it becomes very large. So, even though there is no obvious source for sound in this setup, if the amplifier is set high enough, it will spontaneously start making a sound, which you hear as feedback.

What determines how large the signal will get, as it clearly cannot grow indefinitely? An amplifier, like the amplifier in your stereo system, takes a weak input signal and makes it stronger so that it can drive speakers. Amplifiers increase the input signal by a constant factor, called the gain. In the examples above, we considered gains of 10 and 200. The only problem comes when the input signal is too large. The amplifier tries to increase this signal by its gain factor. However, there is a maximum output signal that the amplifier can produce. Bigger amplifiers can produce a bigger output signal, but every amplifier has some limit. Once this limit is reached, the amplifier cannot produce a bigger output signal. This is called saturation. In the second feedback loop that we considered above, the signal got bigger on each pass through the loop. However, at some point, the amplifier stops increasing the signal, and the signal reaches a constant level. When the circuit is turned on, the signal rises rapidly, until the amplifier reaches saturation, at which point the signal stays constant.

Although the example may seem somewhat complex and technical, there are a couple of general principles that apply to all feedback loops. In any loop, there is loss. In the example above, the signal can become weaker because all of the sound from the speaker does not go to the microphone (loss in the tube). There is also loss because the microphone does not turn all of the sound it receives into electrical energy. Loss *decreases* the signal by a constant factor. To compensate for this, feedback loops require a source of gain. In the example above, the audio amplifier provides this gain. The amplifier *increases* the signal by a constant factor, but only up to a point when the amplifier saturates.

For feedback to occur, the gain must be greater than the loss. For example, the loss may be a factor of 100—that means that the signal is reduced by a factor of 100. If the gain is 10, the amplifier partially makes up for the loss, but not completely, and as we saw in the example above, there will be no signal. However, if we increase the gain by turning up the volume on the amplifier to 200, the gain now is greater than the loss. The signal is decreased by a factor of 100 by the loss and increased by a factor of 200 by the gain, resulting in an overall gain that is 200/100 = 2. It is this overall gain that determines what the feedback loop will do. If the overall gain is greater than 1, feedback will occur. If it is less than 1, no sound will be produced. The gain on the amplifier affects this overall gain. If we start with the volume very low, the overall gain will be much less than 1 and nothing will happen. If we start turning up the amplifier, the overall gain will get bigger, but as long as it is less than 1, nothing will happen. At some critical point, as we keep turning up the volume, the overall gain will become greater than 1 and suddenly feedback will occur and we will hear sound. This is a very sharp threshold, as the overall gain has to simply go from just less than 1 to just barely greater than 1. Feedback is demonstrated in Lab 5.1.

So far, we have determined the conditions that will produce feedback—there must be a feedback loop and the overall gain of the loop must exceed 1. But, an equally important question is what will be the frequency of the feedback? In Chapter 3, we saw that an open-open tube can be described by a series of overtones, or resonances. If the frequency applied to the cavity is equal to a resonant frequency, the microphone will produce a larger output signal, compared to some random frequency. To put this in the context of this section, a frequency equal to the resonance of a cavity will have the smallest loss. As a result, an overtone will have the largest overall gain. So, in our experiment where we start with the volume very low and slowly turn it up, the first frequency to show feedback will be an overtone (or the fundamental) of the cavity. Once feedback starts on a particular frequency or overtone, the feedback will generally continue on that frequency. This is because, once the amplifier has saturated, it loses its ability to amplify other frequencies, as well. In this way, feedback at one frequency suppresses all other frequencies.

There is one final consideration for determining which overtone, or cavity mode will produce feedback. The microphone cannot sit on a node of the cavity, as there will be no signal, regardless of the gain. This is actually useful if one is trying to select which mode will show feedback. The conditions for feedback in an acoustic cavity are summarized as follows:

1. The overall gain must be greater than one.
2. Feedback will occur on a cavity mode.
3. The speaker and microphone cannot be on a node for a mode to show feedback.

These ideas can be tested in Lab 5.1.

Worked Example

Consider an amplifier and an open-open acoustic cavity with the following gain profile:

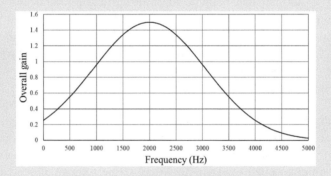

A. Over what range of frequencies is feedback possible?
B. If the tube is 13.7 cm long, which of its modes could show feedback?
C. Where can we place a microphone to favor the lowest mode?

Solution

A. The overall gain exceeds 1 in the range of 1000–3000 Hz.
B. The fundamental frequency of an open-open tube 13.7 cm long is $f = 1{,}252$ Hz. The next mode has a frequency of 2,504 Hz. These are the only modes that will show feedback.
C. The gain for the second mode is greater than for the first mode, so we need to prevent feedback on the second mode. The second mode has a node in the middle of the tube. If we place the microphone there, we will only get feedback on the lowest mode.

As we have seen in this section, feedback is a rather complex phenomenon and it may not be so clear why we are studying it or what use it has. The first step is to just try to get a sense of what feedback is. In the next section, we will consider feedback in a much broader context and show examples of the many important applications of feedback.

5.2 Applications of Feedback

In the previous section, we introduced a very specific example of feedback in an acoustic cavity. Actually, there are many examples of feedback and they fall into two categories: positive and negative feedback. In any feedback loop, the output of the loop goes around to modify the input to the loop. Of course, since it is a loop, there really is no separate input or output. When the loop modifies the signal, it can either exaggerate the signal or minimize the signal. The first case is positive feedback and the second case is negative feedback.

Examples of Positive Feedback

The acoustic cavity discussed in the last section is a perfect example of positive feedback. Above, we stated that positive feedback exaggerates the signal. Let's consider the case of acoustic feedback where the overall gain is say 1.5, and we start with a signal of 0.1 V. After one loop, the signal will be 0.15 V, and in the next loop, the signal is 0.225 V. Thus, if the signal is getting bigger, it continues to get bigger. In other words, it exaggerates the signal. Now, if the gain is 0.5, the signal will be 0.05 V after one loop, and 0.025 V after the next loop. Here if the signal is getting smaller, it continues to get smaller—again, the loop exaggerates the effect. This is very similar to the unstable equilibrium we discussed in Chapter 4, Fig. 4.2. Here, a slight displacement from equilibrium leads to an ever-increasing displacement—exaggerating the initial displacement.

One other example of positive feedback is explosions—either chemical or nuclear. In any explosion, energy stored in a molecule or a nucleus is released. However, the energy stored in any signal molecule or nucleus is very small. What is needed for an explosion is for all the particles to release their energy at nearly the same time. This is done through feedback. The idea is that if one particle releases its energy, it makes it more likely that other particles will release their energy, as well. In a chemical reaction, this occurs because the temperature starts to increase and the increasing energy makes it more likely that other molecules will release their energy. Of course, this further increases the temperature, making it even more likely that other molecules will give up their energy, etc. When a nucleus releases energy it gives off neutrons. However, a neutron can encourage another nuclei to release energy and more neutrons. Thus, the number of neutrons starts to increase, making more likely that more nuclei will give up more neutrons, etc. This feedback can occur very quickly, leading to an explosion.

Examples of Negative Feedback

While a system with positive feedback can change very quickly (i.e., the amount of sound in the cavity, the temperature of a sample, or the number of neutrons) a system with negative feedback does not like to change, and this is a very important property. For example, one wants to keep the temperature of their house constant. This is done through negative feedback. The loop consists of the following elements:

- A thermometer—to measure the house temperature
- A thermostat—to control a furnace or air conditioner based on the thermometer
- A furnace and air conditioner—to heat and cool air for the house
- The air in the house, which affects the temperature.

What happens if the temperature of the air starts to fall? The thermometer will measure this and tell the thermostat to switch on the furnace. The furnace will then heat the air and reverse the fall in the temperature. If the temperature rises above a certain level, this will affect the thermometer, which will affect the thermostat which will tell the air conditioner to turn on, which will reduce the temperature. In other words, the response of a negative feedback loop is opposite to the change: in the example above, if the temperature rises, the system responds by reducing the temperature. If the temperature falls, the system responds by increasing the temperature. Thus, negative feedback loops are a way of maintaining a system in a constant state. Again, this is similar to the stable equilibrium discussed in Chapter 4, Fig. 4.1. Here, if there is a small displacement, the system brings the marble back towards the equilibrium point, counteracting the initial displacement.

What would happen if the heating system in your house had positive feedback? If the temperature started to rise, the furnace would turn on, further increasing the temperature. If the temperature fell, the air conditioner would turn on further decreasing the temperature. Thus, positive feedback creates very unstable conditions, while negative feedback creates stable conditions.

Another amazing example of negative feedback is the temperature of the human body. We can expose our body to very hot weather (over 100°F) or very cold temperatures (under 0°F) but our internal temperature is maintained at 98.6°F within a few tenths of a degree. Of course, under extreme conditions, the body cannot maintain this indefinitely, but it generally controls this temperature very precisely. This can only be done through negative feedback.

Frequency Sources and Standards

Applying positive feedback to the acoustic cavity discussed above has a remarkable and important application. In Fig. 5.1, we used a sound generator to measure the resonant modes of a cavity by sending in known frequencies and monitoring when there was a strong response. However, this raises the question of where do frequencies come from in the first place? We assumed that the sound generator knows how to produce a given frequency, but now we are asking how? Fig. 5.2 provides the answer. In this case, the setup with a speaker, microphone, amplifier and acoustic cavity spontaneously produces a frequency without any external reference. However, since we know how to calculate the overtones of an acoustic cavity from basic physics principles, we now have a method for producing a particular frequency. First, you design a cavity where the desired frequency is the fundamental or overtone of the cavity. Then, you put the cavity into a feedback loop with sufficiently high gain. Finally, you need to ensure that the feedback picks the mode that you want. This can be done in various ways, including placing the speaker and microphone on nodes of the unwanted modes. Alternatively, one can suppress unwanted frequencies using a device called a filter that allows certain frequencies to pass through the system and others to be blocked. In fact, the amplifier in the feedback loop can often perform this task.

Quartz crystal watches work in the same way as the example of the open-open tube—although here, the cavity is the quartz crystal and the sound waves are in the crystal. Even the most precise clocks ever invented, atomic clocks, share the same basic characteristics, except in this case atoms serve as microscopic cavities.

This is also how musical instruments work. Essentially, an instrument faces the same problem as a clock. On an instrument, you want to play a note of a certain pitch. That pitch corresponds to a frequency—but how do you get the particular frequency that you want? You must build a structure (the instrument) that has cavity modes that correspond to the frequency that you want. This can be done with the modes of a vibrating string or an air column. Then, there must be some feedback loop to set that particular mode vibrating. The contact point between the bow and the string in a violin provides that feedback, while the reeds in an oboe or clarinet, or your lips for a flute or brass instrument provide the feedback.

One problem common to all frequency sources is that they can be affected by external conditions, such as temperature. In an acoustic cavity, the speed of sound depends on temperature that can affect the frequency of the cavity modes. Of course, feedback can also solve this problem—with negative feedback and a heater and cooler, the cavity can be maintained at a constant temperature.

5.3   The Doppler Effect

So far, we have only considered stationary sources of sound and stationary listeners or observers. However, if either the source or the observer is moving, things change. This is called the Doppler effect. Like the idea of feedback, covered in the last two sections, the Doppler effect has many important applications. Because the Doppler effect depends on things moving, it can generally be used to determine the motion or speed of an object. Objects of interest may be the speed of a car on the highway, the motion of blood flowing through an artery, the rotation of a galaxy, even the expansion of the Universe. As with many fundamental principles in physics, the range of applications can be truly enormous. The rotation rate of galaxies opens up a very interesting problem: they rotate faster than expected, based on our current understanding of gravity and the makeup of galaxies. Either galaxies contain much more mass than we can see, a hypothesis called "dark matter"; or, Einstein's theory of General Relativity is incomplete. In either case, we currently don't understand a major aspect of our Universe—based on Doppler effect measurements.

So, what is the Doppler effect? One of the most common examples is that of the pitch of a siren on an ambulance or a fire engine. You may have noticed that as a fast-moving siren passes by you, the pitch of the siren abruptly drops in pitch. At first, the siren is coming towards you, when the pitch is higher. After passing you, the siren is going away from you and the pitch is lower. This is a manifestation of the Doppler effect.

There are two different situations for the Doppler effect that we will investigate. The first is when the observer is moving. For example, you are in a moving car and are passing by a stationary siren. In the other case, you are stationary, and the source is moving past you. While the second is perhaps the more common situation, the first is easier to analyze. You also might think that these two situations are the same. As it turns out, they are not and this means that you can also learn about who is moving, the source or the observer. We will return to this question in the next section.

Moving Observer

To understand the moving observer, imagine you are in a motorboat on the ocean, as shown in Fig. 5.3.

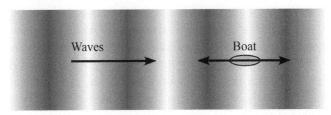

Figure 5.3 Example of the Doppler shift with a moving observer.

If you are not moving, the boat will bob up and down with a certain frequency determined by the ocean waves coming in. However, imagine that you are moving into the waves fairly quickly. You will find that you bob up and down more rapidly, because you hit the crests of the waves sooner than if you were not moving. So, the frequency of the waves appears to be higher to you than if you were not moving. Notice, the waves themselves have not changed, only your experience of them. Nevertheless, you would say that the frequency has increased. Now imagine that you are returning to shore, and so you are traveling in the same direction as the waves. In this case, the waves may still overtake you, but at a much slower rate—you will bob up and down more slowly. In fact, if you travel with exactly the same speed as the waves, you will not bob up and down at all.

The same thing is true for sound waves, or any other waves. If you are moving into a wave, its frequency will appear to you to be higher, while if you are traveling in the same direction as the waves, their frequency will appear to be lower.

The formula for the frequency that the moving observer will detect depends on the speed of the observer—the larger the speed the greater the effect. If we call the speed of the observer, V_o, the frequency the observer detects will be:

$$f_o' = f\left(1 + \frac{V_o}{V_{wave}}\right). \tag{5.1}$$

Here, f is the original frequency and V_{wave} is the speed of the wave. However, above, we saw that the Doppler effect depends on the *direction* in which the observer is moving. How does that enter into this formula? If the observer is moving towards the source of the sound the frequency should go up. That is what the formula predicts—so far so good. If the observer is moving away from the source, the frequency should go down. How can we make this happen? There are two ways to understand this. We can say that if the observer is moving towards the source, its velocity is positive, or greater than zero, while if it is moving away from the source, its velocity is negative, or less than zero. If you put a negative number for V_o into the formula above, the result will be that the frequency decreases. Alternatively, we can write two formulas, one for the observer moving towards the source, and one for moving away from the source:

The observer moving towards the source:

$$f'_o = f\left(1 + \frac{V_o}{V_{wave}}\right).$$
(5.2a)

The observer moving away from the source:

$$f'_o = f\left(1 - \frac{V_o}{V_{wave}}\right).$$
(5.2b)

Notice that there is just a change in sign.

What happens if you are moving away from the source, with a speed, V_o equal to the speed of the wave, V_{wave}? In this case, we would find that $f' = 0$. What does this mean? This is just the case where you are moving along with the waves, and so you don't see the waves going up and down, at all, so there is no frequency to the waves that you are aware of.

Moving Source

The situation where the source is moving is actually a bit more difficult to picture. In the following diagram, Fig. 5.4, the source is marked with a star and is moving to the right. At periodic points in time, it sends off a circular wave. However, once the wave leaves the source, it is no longer affected by the motion of the source—the wave just travels on its own. However, when the source sends off the next wave, it will have moved forward a bit. The new wave is a circular wave, just like the previous one, but its center is shifted slightly in the direction that the source is moving. Figure 5.4 shows the resulting pattern.

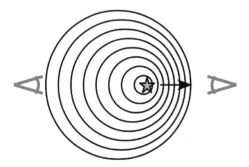

Figure 5.4 Doppler shift with moving source.

Now imagine you are standing on the right with the source coming towards you. You can see that the waves are compressed together. The distance between the crests is the wavelength and so the waves you see will have a shorter wavelength. A shorter wavelength will have a higher frequency. So, if the source is moving towards you, the frequency of the waves will be higher. If you are standing on the left with the source going away from you, just the opposite is true: the waves are spread out, so the wavelength is longer, and the frequency is lower.

As before, we can write down a formula for the frequency detected by the observer, if we call the velocity of the source V_s:

The source moving towards the observer:

$$f_s' = f\left(\frac{V_{wave}}{V_{wave} - V_s}\right).$$ (5.3a)

The source moving away from the observer:

$$f_s' = f\left(\frac{V_{wave}}{V_{wave} + V_s}\right).$$ (5.3b)

Notice that in both cases, moving observer and moving source, if the source and the observer are moving together, the frequency goes up. If they are moving apart, the frequency goes down. However, the amount of the frequency change depends on whether it is the source or the observer that is moving. It is possible to measure the Doppler Effect with earbuds and a microphone, as demonstrated in Lab 5.2.

5.4   Sonic Booms and Shock Waves

The basic formulas for the Doppler shift are given in the previous section. However, there are some consequences and uses of the formulas that are not so apparent. We will discuss some of these in this section.

Sonic Booms

Let's consider the situation where the source is moving towards the observer and think about the consequences of Eq. 5.3a.

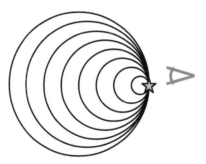

Figure 5.5 Creation of a shockwave.

If $V_s = 0$, which means the source is not moving, the formula gives $f' = f$. In other words, the frequency stays the same and there is no Doppler shift. If the source starts moving, the denominator will get smaller, making the ratio bigger and the Doppler shifted frequency, f', will increase. What happens as the source velocity gets bigger and bigger? The difference $V_{\text{wave}} - V_s$ gets smaller and smaller and f' gets larger and larger. Now, at some point, the source velocity will exceed the wave velocity. Right at this point $V_{wave} - V_s = 0$, making f' infinite. To get an idea of what this means, Fig. 5.5 shows a picture of the waves when $V_s = V_{wave}$.

Because the source is now traveling at the same velocity as the waves, the waves never get ahead of the source, and the waves all pile up at one point. The distance between the waves goes to zero and so the frequency becomes very high. More importantly, all of the energy gets concentrated into a very small distance and this is called a shock wave. In this case, the observer does not hear the approaching source at all until the shock waves hit with all of the energy in the wave. For sound waves, this can cause a very loud noise, called a sonic boom. Any time a source equals or exceeds the speed of the wave, a shock wave will be formed.

If the source is traveling faster than the waves, the waves never catch up to the source, and a different sort of pattern is formed, shown in Fig. 5.6.

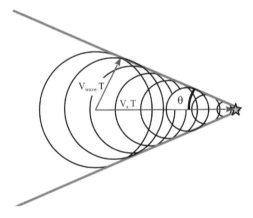

Figure 5.6 Mach cone. The arrows show the distance traveled by the source, V_s, and the wave, V_{wave}, in a time T. From this, the opening angle (θ) can be determined, Eq. 5.4.

The shape of the shock wave is called a Mach cone and the opening angle of the cone is given by:

$$\sin\theta = V_{wave}/V_s. \tag{5.4}$$

Each kind of wave has its own example of shock waves and Mach cones.

Water waves

The easiest waves to visualize are water waves. In this case, you just need to drive a boat faster than the speed of water waves, which is not hard to do (Fig. 5.7).

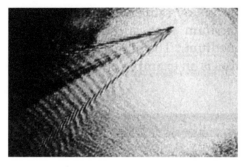

Figure 5.7 Shock wave in water.

The Mach cone is even better in a controlled experiment (Fig. 5.8).

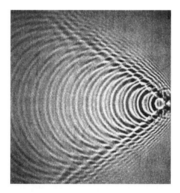

Figure 5.8 Shock wave created in a special ripple tank.

Sound Waves

Sound waves are harder to visualize, but they are easy to hear. If a jet exceeds the speed of sound, it produces a loud sonic boom, which can even rattle or break windows. Jets are not allowed to fly faster than the speed of sound over populated areas.

However, with some sophisticated photography, the Mach cone created by a jet airplane in air can be seen (Fig. 5.9).

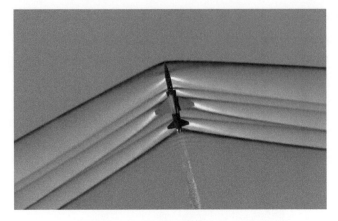

Figure 5.9 Shock wave in air produced by a jet airplane.

A few years ago, a car actually went fast enough to break the sound barrier. In this case, the shock wave kicked up dust as it went past. At http://www.flymicro.com/photolib/ you can see the shock wave alongside the car.

Light Waves

Can you produce a shock wave of light? You might answer no, because to create a shock wave, something must travel faster than the wave. Since nothing can travel faster than the speed of light, this could never happen. However, this is not quite true. In a transparent material, like glass, light slows down by a factor called the "index of refraction". The index of refraction of most types of glass is about 1.5. So, inside of a piece of glass, light travels about 67% of its normal speed. This means that a very fast particle can actually exceed the speed of light in a material. If you accelerate electrons to a very high velocity and fire them into a piece of glass or plastic, they will produce a shock wave of light that you can see. This light is called "Cherenkov radiation".

The following picture shows a nuclear reactor. In the reactor, uranium atoms give off energetic neutrons. These neutrons collide with electrons and give them very high velocities. The entire reactor is underwater to keep it cool. In the water, the speed of light is reduced, just like in glass. The electrons then produce Cherenkov radiation, which appears as a bluish glow (Fig. 5.10).

Figure 5.10 Shock wave of light created by electrons traveling in water: Glow produced by Cherenkov radiation in the Waste Encapsulation and Storage Facility at the Hanford weapons complex.

Reference Frames and Double Doppler Shifts

One final issue needs to be discussed with respect to the Doppler effect. We noted above that if the source and the observer were coming towards each other, the frequency of a tone would increase. However, if we look at the formulas for the moving source and moving observer, the *amount* of the increase will be different depending on whether the source or the observer is moving.

Now, it may seem like a simple matter to determine which is moving, the source or the observer. However, it is not so simple. Say that you are standing by the side of a road, listening for a siren. You would say that you are stationary, and it an ambulance passes by, that is a case of a moving source.

Although you are just standing by the side of the road, you are also standing on the Earth, which is spinning. It is also orbiting the Sun at a very high velocity. Not only that, the Sun is orbiting the center of the galaxy, again, with a high velocity. So, are you moving or stationary? Ultimately, any question relating to whether or not you are moving, must specify *with respect to what?* If I am standing in the classroom, I am not moving with respect to the desk. However, I am moving with respect to the Sun. The system by which you are judging the motion is called the "frame of reference". If you are sitting in a car traveling at 60 miles/h and you are holding a coffee cup, you don't think of the cup as moving, because the obvious frame of reference is your hand. With respect to the ground, the cup is traveling at 60 miles/h, but that does not affect your ability to drink from the cup.

Worked Example

During lift-off, a rocket produces a shock wave as shown:

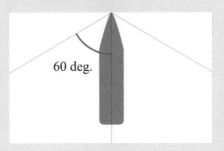

How fast is the rocket moving? Assume 300 m/s for the speed of sound, as the speed of sound decreases with altitude.

Solution

Equation 5.4 tells us that $\sin \theta = V_{wave}/V_s$. In this example, $\theta = 60°$ and $V_{wave} = 300$ m/s. So, $V_s = V_{wave}/\sin\theta = (300 \text{ m/s})/\sin(60°) = 346$ m/s.

So, now, an interesting question arises: when it comes to the Doppler effect and the difference between a moving source and a moving observer, what frame of reference is used to determine which is moving? Actually, there is a very precise answer to this question: since we are talking about waves, there must always be a medium present for the waves to travel in and, so, *the medium is the frame of reference* for measuring all motion.

For example, if you are standing outside and the air is still, you are not moving with respect to the medium for sound (which is the air). If the wind starts blowing at 20 miles/h, you are no longer stationary—you are actually moving at 20 miles/h with respect to the air. If a car is traveling at 20 miles/h in the same direction that the wind is blowing, the car is actually not moving with respect to the air.

As another example, consider a sailboat tied up at a dock, again with the wind blowing at 20 miles/h. You feel the wind blowing against you, so you must be moving with respect to the air. Now, if you are sailing downwind, the boat is being carried along at 20 miles/h by the wind. Under those conditions, the boat feels quite calm—you don't feel any wind. So, in this case, you are now stationary with respect to the air. Figure 5.11 shows how this affects the Doppler shift.

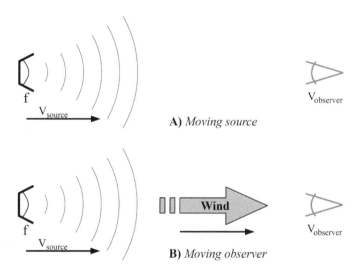

Figure 5.11 Doppler Shift with a stationary and moving medium.

Double Doppler Shifts

Understanding reference frames allows us to calculate a more difficult problem: what is the Doppler shift if both the source and the observer are moving? This would be called a double Doppler shift.

In any problem where both source and observer are moving, the solution involves breaking the problem down into two pieces. First, calculate the Doppler shift for an intermediate observer at rest in the medium. This just requires applying the formula for a moving source, because the intermediate observer is not moving. Once this intermediate frequency, f', is determined, consider the intermediate observer as a source, at rest, sending out the intermediate frequency. This is now a problem of a moving observer, since the (intermediate) source is at rest in the medium. It is a little tricky to work this out, but it just requires two steps instead of one. Figure 5.12 shows the general idea.

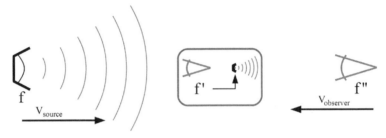

Figure 5.12 Doppler Shift with moving source and observer.

In this example, for the first step, you would use the case of moving source moving towards the observer to find f', while the second step would be moving observer moving towards the source:

$$f' = f\left(\frac{1}{1 - \frac{v_{source}}{v_{wave}}}\right); f'' = f'\left(1 + \frac{v_{observer}}{v_{wave}}\right).$$

Doppler Shifts through Reflection

One of the most common applications of Doppler shifts is in determining the speed of a moving object, like a car. In fact, bats use this technique to determine the speed of a bug that it wants to catch.

Let's say that the air is still and a bat is sitting still on a branch and it hears a bug. The bat will send out a high-pitched tone. This tone bounces off the bug and the bat hears the reflected tone.

If the bug is not moving, the reflected tone will have the same frequency as the emitted tone. What happens if the bug is moving? As with the double Doppler shift above, we consider the problem in two parts.

First, find out what frequency the bug will hear. This would be a case of a moving observer, as the bat is sitting still. This is the intermediate frequency.

Second, consider the bug to be the source of the intermediate frequency and calculate what the bat would hear. The second part would be a moving source, as the bug acts as the source of the intermediate frequency while the bat is now the observer and is not moving. See Fig. 5.13.

Figure 5.13 Doppler shift in reflection.

Again, this may seem a bit complex, but this is really one way that bats can determine the speed of a bug that it wants to catch. In fact, the bat does more. The shift in frequency is usually rather small, as bugs do not move too fast. Rather than try to sense the small shift in frequency, the bat focuses on the beat frequency between the sound that it produced and the reflected sound.

$$f' = f\left(1 - \frac{v_{bug}}{v_{wave}}\right); f'' = f'\left(\frac{1}{1 + \frac{v_{bug}}{v_{wave}}}\right)$$

For velocities much less that the speed of the wave, this can be simplified considerably to: $f_{beat} = 2fv_{bug}/v_{wave}$.

For example, let's say the bat produces a tone at 15,000 Hz and a bug is flying away at 1 m/sec. For the first part, we determine the frequency that the bug will hear. This is a moving observer, and the bug and bat are moving apart, so the frequency will go down. The intermediate frequency will be (15,000 Hz) (1 – (1 m/s)/(343 m/s)) = 14,956 Hz. In the second part, this intermediate frequency is shifted because the bug is now the source and is moving. So, the final frequency that the bat hears in reflection will be (14,956 Hz)(343 m/s /(343 m/s +1 m/s)) = 14,913 Hz. So, rather than try to detect that change in frequency, the bat listens for the beat frequency between the original 15,000 Hz and the reflected 14,913 Hz. This beat frequency is 87 Hz and is much easier to detect. This agrees with the approximate expression above for the beat frequency.

Radar guns used to measure the speed of cars work exactly the same way but with microwaves. Since cars travel *much* slower than the speed of light, the total shift in frequency is tiny. However, the beat frequency is easily detected.

Light and the Theory of Relativity

One final note: the Doppler shift of light is a little different than for other waves. Why this is the case is not well understood, but it has to do with frames of reference. Above, we noted that there is a difference between a moving source and a moving observer and motion was defined relative to the medium. By carefully measuring a Doppler shift under the correct conditions, one can determine one's speed relative to the medium. The medium for light is not well understood, but it appears to be the vacuum, or empty space. However, it was realized that through Doppler shifts you can determine whether you are moving with respect to the medium, and so, scientists got excited because they thought they could learn something about the medium for light, or at least how fast we are moving through it. However, when an experiment was attempted to measure our motion with respect to the medium, it failed. It turns out that there is no difference between a moving source and a moving observer when it comes to the Doppler shift of light.

Albert Einstein eventually figured out why this is and called the answer to this the "Special Theory of Relativity". For this course, we will use the same equations for the Doppler shift of light as for other waves. However, this is another case where light behaves in a peculiar way.

5.5 Problems

Section 5.1

5.1 You are using an amplifier with a gain curve given below to establish feedback in a tube.

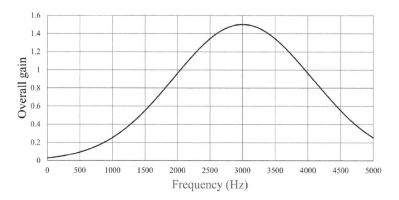

(a) Over which range of frequencies can this produce feedback?

A. 0–5000 Hz. B. 1000–5000 Hz. C. 2000–3900 Hz.
D. 2500–4350 Hz.

(b) Which length of an open-open tube will show feedback in its fundamental mode?

A. 5 cm. B. 10 cm. C. 20 cm. D. 40 cm

5.2 Assume that we have a closed-open air column with a length of 30 cm. We connect a speaker and microphone together through an amplifier with the following overall gain profile:

(a) What are the frequencies of the modes which could exhibit feedback?

(b) Which mode will dominate?

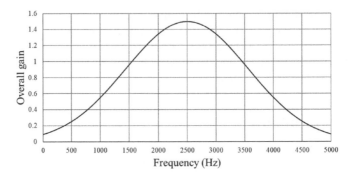

5.3 (a) If we connect a speaker and microphone together through an amplifier with the following overall gain profile, over what frequency range can we get feedback?

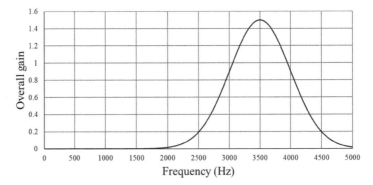

(b) Now, consider that we have a closed-closed air column. If we only want one mode to exhibit feedback we want the frequency interval between modes to equal the frequency range from part (a). What is the maximum length air column that will only have one mode in this range? (Note, this is a rather tricky problem!)

5.4 An amplifier can provide gain only up to a certain frequency:

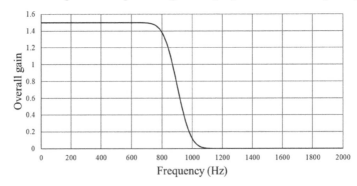

(a) Below what frequency can feedback occur?

(b) If you use a closed-closed air column as a resonator, what is the longest length you can use and only have one mode experience feedback? Hint: you want $2f$ to just avoid going into feedback.

5.5 The following is a plot of the gain of an amplifier as a function of frequency:

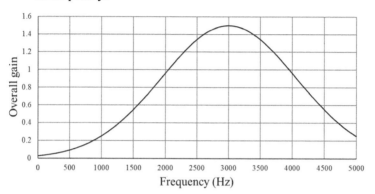

(a) Over what frequency range can feedback occur?

(b) If this were the gain in a closed-closed air column, over what range of lengths of the air column would both the second harmonic, $2f$ and the third harmonic $3f$ both experience feedback?

5.6 (a) You have a closed-closed air column with a length of 34.3 cm. The graph above gives the overall gain in a feedback loop. At what frequencies will you have feedback? (There are 2.)

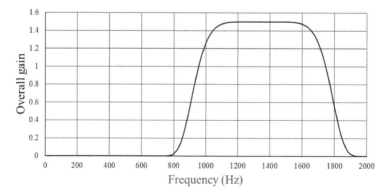

(b) Where along the tube would you place the microphone to make sure that you get feedback on the lower of the two frequencies?

(c) Where along the tube would you place the microphone to make sure that you get feedback on the higher of the two frequencies?

5.7 An amplifier can provide gain only up to a certain frequency:

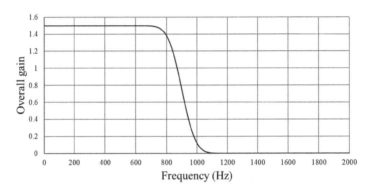

(a) Below what frequency can feedback occur?

(b) If you use a closed-closed air column with a length of 17 cm as a resonator, will the fundamental mode show feedback?

(c) If yes, should you lengthen or shorten the tube to prevent feedback?

If no, should you lengthen or shorten the tube to get feedback?

5.8 (a) Sketch individual modes and the nodal lines for an air cavity, closed all sides, in the mode consisting of the fourth mode in the horizontal direction and the second mode in the vertical direction.

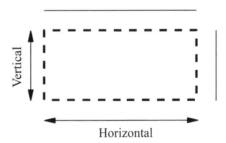

(b) If the horizontal dimension is 1.10 m and the vertical dimension is .37 m, what is the frequency of the mode described in part (a)?

(c) If I created a feedback loop with this cavity, mark a position on the rectangle above where I could place a microphone to ensure that I get feedback on the horizontal mode but not the vertical.

5.9 This rectangular cavity has dimensions of 18 cm by 10 cm.

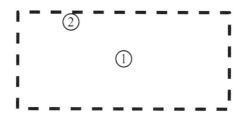

(a) First assume the microphone is placed at the position marked (1). Find the frequencies of the first three modes that can show feedback. You will need to think about where the nodal lines for each mode are.

(b) Now assume the microphone is placed at the position marked (2). Again, find the frequencies of the first three modes that can show feedback.

Section 5.2

5.10 Consider the following scenarios:

(1) Arctic ice melts; less ice reflects less sunlight; less sunlight reflected heats Earth more.

(2) Rabbit population grows; more rabbits means more foxes; more foxes means more rabbits hunted.

(3) Price of cell phone increases; higher price means fewer customers; fewer customers means higher prices.

(4) Blood sugar levels increase; increased levels lead to the release of insulin; insulin converts blood sugar into fatty acids.

Which one of the following statements is correct?

A. 1 and 2 are positive feedback, 3 and 4 are negative.

B. 1 and 3 are positive feedback, 2 and 4 are negative.
C. All are positive feedback.
D. 2 and 3 are negative feedback, 1 and 4 are neither.
E. 1, 2, and 3 are positive, 4 is negative.

Section 5.3

5.11 A siren on a fire truck operates at 960 Hz. If the truck is stationary and you are driving towards the truck at 69 miles/h, what frequency do you hear? Use 767 miles/h for the speed of sound.

5.12 A siren on a fire truck operates at 1060 Hz. If you are stationary and the truck is driving towards you at 67 miles/h, what frequency do you hear? Use 767 miles/h for the speed of sound.

5.13 Many streetlights contain sodium atoms, which strongly give off yellow light at two wavelengths: 589.00 and 589.59 nm. 1 nm = 10^{-9} m.

(a) What frequencies do these wavelengths correspond to and what is the beat frequency between them?

(b) If a spaceship is approaching the Earth at 5 × 10^6 m/sec, what frequencies would the pilot measure from the streetlights, and what would be the new beat frequency?

5.14 (a) A spaceship is approaching the Earth at 90% of the speed of light, announcing its arrival at a radio frequency of 91.7 MHz. Using the normal formula for Doppler shifts, what frequency would we receive the message at?

(b) Einstein's relativistic formula is:

$$f' = f\frac{\sqrt{1+v/c}}{\sqrt{1-v/c}},$$

where v is the speed of approach and c is the speed of light. Using this formula what frequency would we receive? These frequencies are quite different and can be used to confirm Einstein's theory of relativity.

5.15 (a) A siren on a fire truck operates at 1000 Hz. While the truck is coming at you 66.3 miles/h, what frequency do you hear?

(b) After it goes by, what frequency do you hear?

(c) By how many half steps does the frequency change?

5.16 (a) You are in a truck traveling at 40 miles/h and an ambulance with a siren at 2000 Hz going 70 miles/h passes by you. The air is still. We want to figure what frequency you will hear. Two steps are required to solve this problem. Fill in the following table to set up the problem.

(b) Now that you have set the problem up, calculate the final frequency that you will hear.

	Moving source or observer?	What speed?	Frequency goes up or down?
Step 1			
Step 2			

5.17 (a) Assume you are in a physics lab with a speaker at each end emitting an annoying frequency of 3000 Hz. As you try to escape the lab at 3 m/s one frequency will be shifted up and the other down. By how much will they be shifted?

(b) What will be the beat frequency between the two frequencies that you hear?

(c) These speakers form a standing wave. What is the separation between the nodes?

(d) If you are moving at 3 m/s, how many nodes will you pass per second?

(e) How does this relate to part (b)?

5.18 The Doppler shift of light is an extremely important tool for astronomers, as they can use it to measure how fast distance objects are moving. Atoms, like hydrogen, emit light at specific wavelengths. Since the Doppler effect changes the frequency of the wave it also changes the wavelength.

(a) Hydrogen emits a particularly strong line at 121.5 nm. What frequency does this correspond to?

(b) If a star is moving away from us at $3 \cdot 10^6$ m/s, what frequency would we measure?

(c) What is the change in wavelength?

5.19 Many times, the Doppler shift is used under conditions where a wave is reflected off of a target (like a police radar detector used to catch speeders or bats looking for an insect). However, calculating the Doppler shift becomes a little more complicated.

(a) A bat is flying towards a wall at 10 m/s and emits a pitch of 10,000 Hz. What frequency would a bug on the wall hear (if it could hear)? Here, the bat is a moving source and the wall is stationary.

(b) The reflected sound will have the same frequency as that heard by the bug, in other words, the Doppler shifted frequency. What frequency would the reflected wave appear to have for the bat? Now, the bat is a moving observer.

(c) The bat actually listens for the beat frequency between the sound it emits and the reflected sound. What would be the beat frequency in this case?

5.20 In order to get a large Doppler shift, the source or the observer must be traveling at a reasonable fraction of the velocity of the wave. However, police officers can use the Doppler shift of microwaves to accurately measure the velocity of a car, which is MUCH less than the speed of light. The precision comes from measuring beats. Police radar uses a frequency of 10^9 Hz.

(a) If the police officer is standing by the side of the road, what is the frequency of the radar that you would detect in your car after you pass by at 60 miles/h = 27 m/s)?

(b) The microwaves reflected by the car now act as a moving source as they travel back to the police officer. What is the frequency of the reflected microwaves that the police officer would detect?

(c) What is the beat frequency between the original radiation and the reflected radiation? This beat frequency can be very accurately measured.

5.21 (a) A trumpet player standing on the side of a lake sounds an A with a frequency of 440 Hz. You are in a motor moving away from the shore at a speed of 13 m/s. With respect to the medium, which is moving, the source or the observer and what frequency do you hear? The air is calm.

    (b) Now the wind is blowing in the same direction and with the same speed as the motorboat. Again, with respect to the medium, which is moving, the source or the observer and what frequency do you hear?

    (c) At this point the wind reverses direction but still has a speed of 13 m/s. Now what frequency do you hear?

5.22 (a) An ambulance is traveling at 70 miles/h and sounds a siren at 2000 Hz. If you are standing by the side of the road and the ambulance is traveling away from you, what frequency do you hear?

    (b) If you are now following the ambulance also at 70 miles/h it is not clear what frequency you will hear. However, you can consider that you are a moving observer hearing the Doppler shifted frequency from part (a). What frequency do you hear?

5.23 An orchestra is playing outside and the wind is blowing. We want to see how this affects the tuning of the instruments. If the oboe player plays an "A" at 440 Hz and the wind is blowing at 20 miles/h what frequency will be heard by the players upwind from the oboe? What frequency will be heard downwind?

5.24 Two fire alarms go off and you find yourself caught in between them. The frequency of the alarms is 850 Hz.

    (a) If you start running towards one to get out of the building at 5.5 m/s, what frequency do you hear from each speaker?

(b) What is the frequency and period of the beat between the two alarms?

(c) When you are right in the middle of the two alarms, there will be constructive interference, so the alarms will sound loud. How far do you have to move to be at the next point where you have constructive interference?

(d) How long will it take you to move this distance at a speed of 5.5 m/s?

(e) How does this time compare to the period of the beat frequency in (b)?

5.25 (a) A stationary siren produces two frequencies, 1050 and 1150 Hz. If you are moving away from the siren at 25 m/s, what beat frequency do you hear?

(b) In part (a), does the beat frequency change depending on whether or not you are moving?

(c) How do these results affect the transmission of music over radio waves? In other words, are the frequencies that you hear while listening to the radio affected by the Doppler shift? Remember that acoustic frequencies are sent as beat frequencies of the radio waves.

5.26 One train is stopped in a station while another train is entering the station at 35 miles/h. Both trains are blowing their horns at 1500 Hz. The air is still.

(a) If you are sitting in the stopped train what beat frequency would you hear between the two horns?

(b) If you are sitting in the moving train, what beat frequency would you hear?

(c) Now the first train leaves the station at 35 miles/h traveling in the opposite direction to the incoming train. What beat frequency would you hear if you were standing on the platform?

5.27 You are sailing between two banks of a lake. The wind is blowing at 12 m/s from left to right and the boat is moving in the same direction at exactly the same speed. A siren is sounding on each bank at the same frequency of 1750 Hz.

(a) On the boat, what frequencies do you hear from the left and right sirens?

(b) What is the beat frequency between these two Doppler shifted frequencies?

Section 5.4

5.28 Imagine you are sitting at the edge of a stream. You dip your toe into the water to create waves and they are felt by a bug floating in the water. With respect to the water, who is the source, who is the observer, who is moving, and who is stationary?

    A. The bug is the source and is moving; you are the observer and are stationary.

    B. The bug is the source and is stationary; you are the observer and are moving.

    C. The bug is the observer and is moving; you are the source and are stationary.

    D. The bug is the observer and is stationary; you are the source and are moving.

5.29 (a) If you are sitting at the edge of a pond and you dip your toe into the water you will create water waves. If a bug is sitting on a leaf in the middle of the pond it will eventually feel the water waves. Who is the source, who is the observer, and is either one moving?

    (b) Now you are sitting at the edge of a stream. Again, you dip your toe into the water to create waves and they are felt by a bug moving along with the water. With respect to the water, who is moving and who is stationary?

    (c) If the water is flowing down the river faster than the speed of the water waves, can you make any waves travel upstream? What is this situation similar to?

5.30 You are sitting at the edge of a river and the water is flowing at 0.75 m/s and the speed of the water waves is 0.3 m/s. A log sticks straight out of the water and ripples form around the log. What angle, θ, will the ripples make?

5.31 (a) Under what conditions can something produce a sonic boom? How fast must it be traveling?

(b) How can you use the effect of a sonic boom to illustrate the difference between a moving source and a moving observer? Use the equations for each to support your answer.

5.32 (a) The pilot of a jet airplane is trying to break the sound barrier. His speedometer always gives his speed with respect to the ground. If there is no wind blowing, how fast must he travel to break the sound barrier?

(b) Now, assume that the wind is blowing at 200 miles/h. If the plane is flying into the wind, how fast, with respect to the ground, must he fly to break the sound barrier? If he is flying with the wind, how fast must he go?

(c) You are at the center of a hurricane and you hear a cow mooing, which is being carried along with the wind. Who is the source, who is the observer, and who is moving?

(d) The storm gets really bad and the winds reach 800 miles/h. If you stay fixed to the ground, would you produce a sonic boom? Why or why not?

5.33 You are in a sailboat sailing downwind. Because the boat is traveling with the wind you feel no breeze even though the wind is blowing at 20 miles/h with respect to the ground. You are moving directly away from the shore where your friend is playing a note on a tuba at 120 Hz. (Note: this problem is about sound waves, not water waves!)

(a) With respect to the medium, who is the source, who is the observer, and who is moving?

(b) What frequency of sound does the observer hear?

5.34 Assume you are in the desert trying to break the sound barrier in a racecar. There are two people, one standing at the start line and one at the finish. You have a siren at a frequency of 2500 Hz.

(a) When you reach a speed equal to one half of the speed of sound, what frequency will the person at the start line hear from the siren?

(b) When you reach a speed equal to one half of the speed of sound, what frequency will the person at the finish line hear from the siren?

(c) If the wind is blowing at 40 miles/h from the start line towards the finish line, how fast would you have to drive with respect to the ground to break the sound barrier?

(d) If the wind is blowing at 40 miles/h from the finish line towards the start line, how fast would you have to drive with respect to the ground to break the sound barrier?

5.35 A fighter jet is going at Mach 3, which is 3 times the speed of sound. What is the opening angle of the cone produced by the sonic boom?

5.36 The jet in Fig. 5.9 is moving in air. How fast is it moving?

5.37 You are looking up in the sky when you notice an airplane (flying from right to left in the figure). Assume that the speed of sound is 767 miles/h.

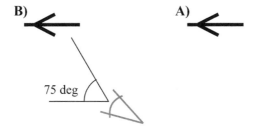

(a) When the airplane is at position (A) what can you tell about the airplane's speed if you can hear it, or if you cannot hear it?

(b) If you cannot hear the airplane until after it flies directly overhead to a position that is at 75° with respect to the horizontal, position (B), how fast is the airplane flying?

Chapter 6

Psychoacoustics: Consonance and Dissonance

6.1 Structure of the Ear and Pitch Perception

Hearing and sound processes start with a complex organ, called the ear. While clearly biological in nature, it utilizes various basic physics principles we have studied throughout this text. The most relevant part of the ear for our purposes is the cochlea and basilar membrane. However, before we discuss those parts, it is useful to get a general understanding of the anatomy of the entire ear structure (Fig. 6.1). We will follow the path of the acoustic energy from where it enters the ear to where the information is encoded in nerve impulses.

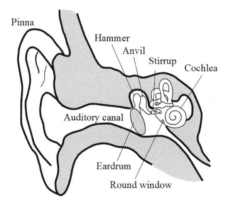

Figure 6.1 Parts of the ear.

The Music of Physics: An Introduction to the Harmonies of Nature
George N. Gibson
Copyright © 2024 Jenny Stanford Publishing Pte. Ltd.
ISBN 978-981-4968-99-7 (Hardcover), 978-1-003-48481-3 (eBook)
www.jennystanford.com

The *earlobe* or pinna is the part of the ear that can be seen from the outside and what is often thought of as the "ear". Its function is to collect the acoustic energy present as sound waves. In addition, the folded shape of the ear lobe concentrates and guides high frequencies deeper into the ear. It also helps give a sense of direction to the source of the sound.

The *auditory canal* is a tube that connects the earlobe to the inner ear region—specifically the eardrum. As a result, it is an open-closed resonator that enhances frequencies around its fundamental resonance, which is in the range of 2000–5000 Hz. Measuring the resonant frequency provides a way to determine the length of the canal, which is sometimes used for fitting hearing aids.

The *eardrum* is a thin membrane closing the auditory canal. It serves to convert sounds waves to mechanical vibrations, much like a microphone.

The *inner ear bones* (hammer, anvil, and stirrup) transfer mechanical energy from the eardrum to the *oval window*, another membrane which can vibrate. The earbones can regulate the amount of energy transferred from the eardrum to the oval window to protect against loud noises. The oval window then sends the acoustic energy into the cochlea.

The *cochlea* is the main structure for processing the acoustic signal received by the ear. It consists of a fluid-filled canal with the basilar membrane running along its length. The canal runs along one side of the basilar membrane and comes back along the opposite side with the acoustic energy following the same path. The whole cochlea is rolled up into a snail-like structure.

The *basilar membrane* acts as a frequency analyzer, in which different parts of the membrane respond to different frequencies. Figure 6.2 shows an uncoiled view of the basilar membrane. At the near end, the membrane is narrow, stiff, and thin, favoring high frequencies and, at the far, end the membrane is wide, loose, and thick, favoring low frequencies. This directly follows our qualitative understanding of the resonant frequency of a vibrating string (Eq. 1.2). Longer lengths, lower tension, and larger thickness produce low frequencies, while shorter lengths, higher tension, and smaller thickness produce high frequencies.

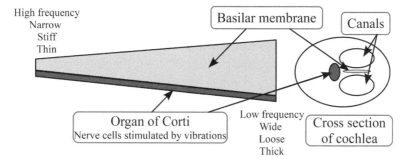

Figure 6.2 Detailed view of the basilar membrane.

How Is Pitch or Frequency Determined?

Sound processing is very complex and quite remarkable. The nominal range of frequencies that humans can hear is from 20 to 20,000 Hz, although age can greatly reduce this range. The range in loudness that the ear can still properly process is an impressive 10^6. For example, hearing a pin drop in a quiet room versus talking in a loud concert. This is due, in part, to the function of the ear bones, regulating how much energy gets transferred to the inner ear, which can both attenuate and amplify the signal. In addition, the nerve cells in the cochlea can provide a degree of positive feedback to provide further amplification. While feedback can enhance sensitivity, it can go too far and start the basilar membrane to vibrate by itself. When this occurs, the ear actually produces a tone and is one cause of tinnitus, or "ringing" in the ear. There are three important aspects of pitch perception:

1. The basilar membrane resonates when a sound wave enters the canal in the cochlea. As discussed above, different parts of the basilar membrane resonate at different frequencies. Thus, the position on the membrane encodes frequency information.

2. The resonant frequencies on the basilar membrane are spaced logarithmically (Fig. 6.3). Compare this to a linear scale which covers the same range of human hearing (20–20,000 Hz). The range of notes on a piano goes from 55 Hz to about 4000 Hz. On the linear scale, this entire

musical range is compressed into just one fifth of the full range. Human speech covers an even smaller range within that (10–300 Hz). While very high frequencies are important, they don't need 80% of the full range. The logarithmic scale covers the different ranges in a more balanced way. Now, the musical range covers about 80% of the full range and human speech falls right in the middle, while still providing space for the very high frequencies. Interestingly, this is the exact same layout as a piano. An interval of an octave covers the same physical distance on the piano, independent of the frequency of the first note. The same is true on the basilar membrane.

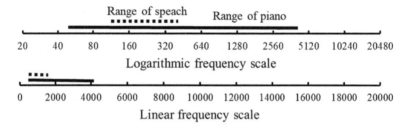

Figure 6.3 Linear vs. logarithmic scales showing full range of human hearing, range of speech (dotted lines), and range of a piano (solid line).

3. A single input frequency can activate a range of nerves on the basilar membrane, shown in Fig. 6.4.

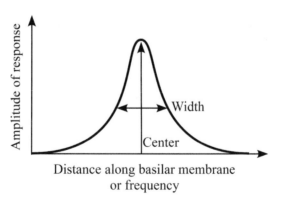

Figure 6.4 Response of the basilar membrane.

This is the response to a pure tone. Such a response is characteristic of any type of resonance and is characterized by the center of the response and the range or width of the response.

6.2   Critical Bandwidth and Just Noticeable Difference

Hearing perception is determined, predominately, by two main parameters, the critical bandwidth and the just noticeable difference. As discussed in the last section, when a pure tone reaches the basilar membrane, the region of the membrane tuned to this frequency responds and activates the nerves at that point, or the center of the response. However, more than one point on the membrane responds, and this range is called the critical bandwidth (CB). In other words, if a tone of 1000 Hz is heard, the area on the membrane tuned to 1000 Hz responds the most, but, in addition, the part of the membrane above and below the center will respond to varying degrees, tracing out the curve shown in Fig. 6.4.

The CB is rather large, about 19% of the center frequency. So, at 1000 Hz, the CB is 190 Hz. Or, the entire part of the membrane tuned to the range of 905 to 1095 Hz will respond. The interval from one end of the response to the other corresponds to about three half steps ($1.0595^3 = 1.19$) or about a minor third.

At 500 Hz, the CB = 95 Hz. Below 500 Hz, the ear is not so good and the CB is relatively constant at about 95 Hz. Notice that none of these values are precise. Most of the response of the basilar membrane occurs right at the input frequency, while there is still some response quite a bit farther out than the CB, but the CB still gives the basic range of the membrane that responds to a particular frequency.

Detecting Pitch

Pitch information is obtained primarily from the place on the basilar membrane that resonates with an input tone. However, some information can be obtained from how often the neurons fire. The neurons fire at the peak of the vibration. They cannot fire fast enough to fire at every peak in the wave. But, when they do fire, it is at the peak. So, it is not quite as simple as saying that the neurons fire at the same frequency of the input tone, by they do fire *in phase* with the input.

Between these two mechanisms, the ability of the ear to identify the exact frequency of an input tone is called the just noticeable difference (JND). Another way to state this is: how much must a frequency change for the ear to notice the change in frequency? It turns out that for the ear, the JND is about 0.5% or 0.005. This is about 1/12 of a half step.

For example, at 1000 Hz, the JND is 5 Hz. So, if two tones are played *separately* at 1000 Hz and 1002 Hz, you would not be able to tell that the pitch has changed. However, if the tones are 1000 and 1010 Hz, you should be able to tell the difference. You can test this in Lab 6.1.

Relationship between the CB and JND

Notice that the CB is much greater than the JND. This is true in general and you can understand why from other experiments.

Here is what the spectrum of a rectangular cavity looks like (Fig. 6.5 produced in Lab. 3.5):

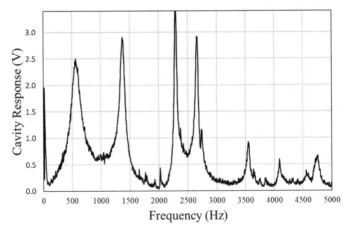

Figure 6.5 Response of a rectangular cavity.

Let's enlarge the *x*-axis around on individual peak (Fig. 6.6).

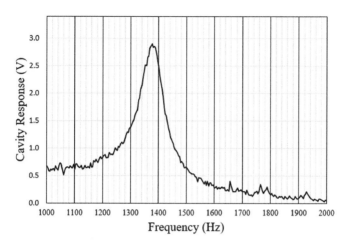

Figure 6.6 Individual resonance peak.

There is a peak at around 1380 Hz. How well can we judge this? The frequency of this peak is certainly greater than 1370 Hz, but less than 1390. So, the frequency is 1380 to within about 10 Hz. This would give a JND of about 0.7%. This is a little worse than the ear can do.

However, notice how there is a range of frequencies that respond, from about 1300 Hz to 1430 Hz. This is about 130 Hz, and this would correspond to the critical bandwidth. In this case, the CB is 10%. This is somewhat less than for the ear, but it is a very different situation.

Actually, the most interesting parameter is the ratio of the JND to the CB. If a peak is very broad, it will be much harder to judge the position of the center. So a larger CB will lead to a larger JND. For our example above, the CB = 10% and the JND = 0.7%, so the ratio of the JND to the CB ratio is 1/14. In other words, we could tell the center frequency with an accuracy of 1/14 of the width of the peak. How does the ear do? In this case, the CB = 19% and the JND = 0.5%, so the ratio is now 1/38. So, given a certain width, the ear is actually very good at determining the center frequency.

6.3   Pure Tones, Consonance and Dissonance

We briefly discussed the perception of consonance and dissonance of pure tones in Section 5.5. We stated that beat frequencies in the range of 15–50 Hz are annoying to the ear. We can now explain this in a bit more detail based on the structure of the ear that we developed in the last two sections.

Two Pure Tones Played Together

Loudness. The first consequence of the CB is on how the loudness of two tones is perceived. If two tones are within the critical bandwidth, the combination does not sound as loud as if they are separated by more than the CB, shown in Fig. 6.7.

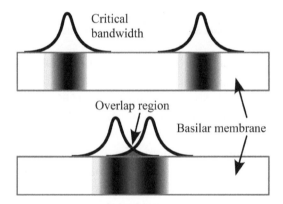

Figure 6.7 Response of the basilar membrane to two pure tones.

If the two tones are quite separate in frequency, as much of the basilar membrane is activated as possible. However, if the CB of the tones overlap, a smaller part of the membrane is activated, and the two tones do not sound as loud. If the neurons are already firing in response to one tone, it does not help to have the other tone also trying to make them fire.

Roughness. If tones are overlapping, there is a region on the basilar membrane where the neurons are trying to respond to both tones. The problem is that the neurons don't know which frequency to follow. The neurons end up jumping back and forth between the two frequencies. This gives rise to a rough or unpleasant sensation. It also makes it hard to focus one's attention on just one of the two tones.

Perception of Two Pure Tones f_1 and f_2

We are now in a position to somewhat quantitatively analyze how two pure tones will be perceived based on the difference of the two frequencies $|f_1 - f_2|$, summarized in Table 6.1.

Table 6.1 How the ear responds to various beat frequencies

$\|f_1 - f_2\|$ < 10 Hz	You hear a single pitch equal to the average frequency ½ $(f_1 + f_2)$. However, you will also hear beats.
> 15 Hz	You will still hear one tone, but the brain can no longer follow the beat frequency. The CBs are overlapping so the neurons fire irregularly creating dissonance. A difference frequency of 30–40 Hz is the worst in terms of dissonance or roughness.
> half step	You can begin to hear two separate tones, but the CB's are still overlapping, so the tones are dissonant.
> CB	You can hear two distinct tones and they sound consonant. You sometimes also hear a difference tone.

Notice how the frequencies of the notes will affect the perception. The first distinctions, 10, 15 and 30–40 Hz are absolute frequencies, independent of the frequencies of the individual notes. The last two conditions (> half step and > CB) are defined relative to the frequencies of the two notes. Thus, these conditions will apply differently for different ranges of frequencies.

For example, consider a perfect fifth between C = 64 Hz and G = 96 Hz. Usually, the perfect fifth is considered to be a consonant interval, even with complex tones. However, the difference between these two frequencies is 32 Hz. This is in the range of dissonance. Thus, normally consonant intervals, when played at very low frequencies will sound dissonant. For this reason, composers rarely use a simple perfect fifth in the low range of a piano. Rather, they will use the interval of an octave and a perfect fifth. Now, the notes would be C = 64 Hz, and G = 192 Hz making the difference frequency 128 Hz, well within the range of consonance.

Difference Tones

Once the beat frequency or difference tone exceeds the lowest frequency that you can easily hear, the brain starts to interpret the beat frequency as a new note. For example, if one plays an A = 220 Hz and an E = 330 Hz, the difference tone is at 110 Hz, and it will sound like another note is being played an octave below the A.

6.4 Complex Tones, Fourier Analysis, and the Missing Fundamental

Complex Tones

So far, we have considered consonance and dissonance between pure tones. Pure tones are simple sine waves at a single frequency. No instrument produces a pure tone, but pure tones are useful in trying to understand how hearing perception works. We found that if two pure tones are separated by less than a critical bandwidth (or about a minor third) the two tones will sound dissonant or rough. If the difference is small enough (<15 Hz) you will hear beats. What was perhaps more surprising is that if the tones were separated by more than a minor third, the tones would always sound consonant. While there are many consonance intervals larger than a minor third (like the perfect fourth and perfect fifth) there are also dissonant intervals (like the seventh and the diminished intervals). How do we explain dissonance of tones that are far apart (i.e., greater than a minor third)?

This was discussed in Section 3.2: the key is to realize that musical tones, as produced from an instrument, are not pure tones. Rather, they consist of a fundamental and harmonics. Thus, when you play A = 440 Hz on a violin, the violin produces a frequency of 440 Hz, but in addition, it is also producing sound at 880 Hz, 1320 Hz, 1760 Hz, ... This is referred to as a complex tone. However, we now know how to compare two complex tones more accurately: we must see if *any pairs of harmonics* lie within a CB. These pairs are analyzed in exactly the same way as pure tones were analyzed in the last section.

For example, let's consider a perfect fifth where the first note is at A = 220 Hz and the second note is at E = 330 Hz. With harmonics, the following frequencies will be present:

220, 440, 660, 880, 1100, 1320, 1540, ...
330, 660, 990, 1320, 1650, 1980, ...

Possible pairs to consider are (220,330), (330, 440), (880, 990), (990, 1100), etc. The last two pairs are separated by less than the CB, but the beat frequency is quite high (>100 Hz), so they will add a small amount of dissonance, but not too much. More importantly, the two notes have a common harmonic at 660 Hz, although this does not cause dissonance, because the difference is less than 10 Hz (in this case the difference is zero).

Now let's consider a diminished sixth. This is a perfect fifth plus a half step. Again, starting at 220 Hz, a diminished sixth will be at $330 \times (1.0595) = 350$ Hz giving:

220, 440, 660, 880, 1100, 1320, 1540, ...
350, 700, 1050, 1400, 1750, 2100, ...

Now consider the pairs (660, 700), (1050, 1100), (1320, 1400). These have beat frequencies of 40 Hz, 50 Hz, and 80 Hz. These are all in the range of roughness or dissonance. Since there are many pairs like this, the overall effect will be dissonant, and, indeed, the diminished sixth is considered to be a dissonant interval. Consonance and dissonance of pure and complex tones are explored in Lab. 6.2.

Fourier Analysis and the Missing Fundamental

The above analysis is sufficient to explain the perception of consonance and dissonance of pure and complex tones. However, it raises an interesting question about complex tones. To create a complex tone, we start with a pure tone at the pitch that we want, as 220 Hz. We then add harmonics to change the tone quality of the note—this can make the note "richer" or "brighter" or more "mellow", any of the qualities we associate with a musical tone. The question is: why does the adding of harmonics not change our perception of the *pitch* of the note—only the tone quality?

The answer to this was provided in the early 1800s by a mathematician/physicist named Fourier. What Fourier realized was that if you add a harmonic to a fundamental, the resulting wave always has the same period as the fundamental, no matter how many harmonics you add! Thus, the resultant wave always seems to have the same pitch as the fundamental, even though is can sound quite different. Fourier also proved the reverse statement: a complex but repeating wave can always be produced by starting with the fundamental frequency and adding harmonics.

So far, this is similar to the discussion in Section 3.2, although in the example below (Fig. 6.8), we are adding 1*f*, 2*f*, and 3*f*. And, again, we see that adding harmonics does not change the period of the complex tone, as all harmonics also repeat with the period of the fundamental.

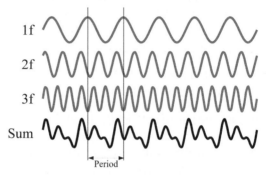

Figure 6.8 Fundaments, harmonics, and their sum.

A very interesting observation follows from this discussion. Because all of the harmonics repeat with the fundamental period, the fundamental frequency does not need to be the loudest to preserve the pitch of the note. In fact, the fundamental need not be present at all. Even if the fundamental is removed from a note, the note still seems to have the same pitch.

Consider the wave above, but without the fundamental (Fig. 6.9).

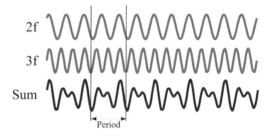

Figure 6.9 Missing fundamental.

Even though the fundamental was not added to the total wave, only 2*f* and 3*f*, the period of the total wave is still equal to that of the fundamental. Of course, if you played 2*f* alone, it would sound an octave higher than the fundamental. But if you add 3*f*, you will hear the pitch drop to the fundamental. This effect is called the "missing fundamental". Strange as it seems, there is an easy way to convince yourself that it must be correct.

Imagine (or find) a piano and play middle A = 440 Hz. Then play the A one octave lower, at 220 Hz. As you listen, you would say that the pitch went down an octave. Now play the A an octave lower, at 110 Hz. Again, the pitch drops by an octave. Go down another octave to 55 Hz, and another to 27.5 Hz. Each time, you will claim to hear the pitch drop by an octave. The interesting thing is that you cannot hear a pitch of 27.5 Hz! Even 55 Hz is very hard to hear. So, although you cannot hear these low pitches, the ear is telling you that the pitch is dropping by an octave. Therefore, the brain perceives a pitch that is not actually present—it fills in the "missing fundamental". In this case, the fundamental is missing because the ear does not respond to such a low frequency.

More technically, the brain finds the greatest common divisor of all the harmonics that it hears. In other words, it looks for that frequency from which all the harmonics can be formed. For example, consider the series: 600, 800, 1000, 1200 Hz... If 600 Hz were considered to be the fundamental, 800 would not be a harmonic of this fundamental, and the overtone series would not make sense to the ear. The largest number that evenly divides each frequency of this series is 200 Hz. Then the series is 3*f*, 4*f*, 5*f*, 6*f*, ..., with *f* = 200 Hz being the fundamental. So, in this case, the "pitch" of the note is 200 Hz. Of course, 100 Hz also divides all of the frequencies in the series, as do 50 Hz and 25 Hz, etc. That is why we must consider the highest frequency that divides all of the overtones.

Lab 6.3 produces the effect of the "missing fundamental" very convincingly.

6.5 Difference Tones

The last topic that we will consider is difference tones. In Section 6.3, we discussed the perception of two pure tones played together. How the brain perceives the pair of notes is controlled by the difference frequency: $|f_1 - f_2|$. If the difference is greater than a critical bandwidth, the pair is perceived as consonant. However, there is still a beat frequency given by $|f_1 - f_2|$. If this beat frequency is high enough, it can be heard as a pitch of its own. For example, if you start at a pitch f_1 and play another note a perfect fifth higher at $f_2 = (3/2)f_1$, the difference frequency between these two notes will be at $(1/2)f_1$ or a pitch one octave lower than f_1. This is actually useful for tuning violin strings: the strings on a violin, viola or cello are all separated by a perfect fifth. If you play two adjacent strings at the same time and they are tuned perfectly, you can hear a difference tone at a pitch one octave below the lower string.

There is not too much more to say about difference frequencies besides the fact that they are real and you can hear them, although it may be rather difficult, for an untrained ear. However, while experimenting with difference tones, I heard a curious effect. When two notes are tuned exactly a perfect fifth apart, I heard a difference tone an octave lower, as discussed above. But, when I detuned the upper note slightly, I heard the difference tone start beating. The question is why did I hear beats in the difference tone, at all, and what two frequencies were beating together?

Previously we considered what happens when a note is tuned away from the ideal case and how this affects the consonance and dissonance of pure and complex tones. To analyze the effect of detuning a note on difference tones, we will say that the frequency is changed by a small amount, called δ. For example, let's consider the perfect fifth with complex tones, and the upper note slightly detuned:

Table 6.2 Beats among difference tones

	First note	Second note
Fundamental	$1f$	$1.5f + \delta$
Harmonics	$2f$	$3f + 2\delta$
	$3f$	$4.5f + 3\delta$

From this table, we can see that the common harmonic at $3f$ will beat with a frequency of $|3f - (3f + 2\delta)| = 2\delta$. In other words, if the upper note is 4 Hz out of tune, you will hear a beat frequency of 8 Hz in the harmonics. But, what about the difference tones? Between the two fundamentals, $1f$ and $1.5f + \delta$, there will be a difference frequency of $0.5f + \delta$. However, between $2f$ and $1.5f + \delta$, there will be a difference frequency of $0.5f - \delta$. So, there are two ways to get a difference frequency an octave lower: one is at $0.5f + \delta$ and the other at $0.5f - \delta$. What frequency will these two notes beat at? We have

$$|(0.5f + \delta) - (0.5f - \delta)| = 2\delta.$$

So, in fact, if I play two notes separated by a perfect fifth, I will hear a difference tone. If I play one note out of tune by say 5 Hz, the difference tone will beat with a frequency of 10 Hz.

This example is here to demonstrate some of the consequences of difference tones and of playing notes slightly out of tune. However, it is also meant to show that how we perceive sound is quite complicated and subtle. In fact, Chapter 6 has only given the most basic elements needed to begin to understand the perception of sound. Indeed, much more is going on inside the ear and brain and this continues to be an area of active research and controversy. Nevertheless, all of these effects are rooted in fundamental physics and waves and ultimately how we perceive sound cannot be separated from the physics of waves.

6.6 Problems

Section 6.1

6.1. Match the following parts of the ear to their function.

(a) Outer ear (pinnae); (b) Ear canal; (c) Eardrum; (d) Ear bones; (e) Cochlea

A. Transfers sound waves from the outer ear to the eardrum

B. Turns sound waves into mechanical vibrations

C. Collects sound waves and helps localize the source

    D. Transfers mechanical vibrations to the inner ear—can protect the inner ear

    E. Primary area where information in sound waves is turned into nerve signals

6.2. Which of these statements is true about the basilar membrane? Points taken off for incorrect answers!

    (a) The basilar membrane is long and narrow and vibrates at only one frequency at a time.

    (b) The basilar membrane is loose and wide at one end; narrow and stiff at the other.

    (c) The wide end of the basilar membrane responds to high frequencies.

    (d) Feedback in the basilar membrane enhances the sensitivity of the ear.

    (e) The resonant frequencies of the basilar membrane are spaced linearly along the membrane.

    (f) The critical bandwidth of the basilar membrane refers to the entire range of frequencies that the ear can hear.

Section 6.2

6.3. Which curve has a width of 50 Hz?

    Which curve has a width of 500 Hz?

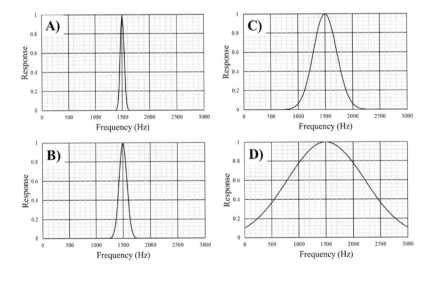

6.4. How accurately can you judge the center of this peak?

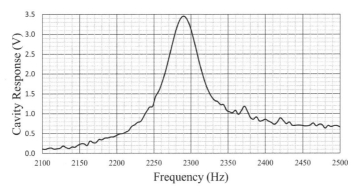

A. 3 Hz

B. 30 Hz

C. 300 Hz

D. 3000 Hz

6.5. What is the width of the peak in Problem 6.4?

A. 7 Hz

B. 70 Hz

C. 700 Hz

D. 7000 Hz

6.6. You can hear the difference in pitch between the perfect fifths according to the just and equal temper scales (i.e., the notes are more than a JND apart.)

(a) True

(b) False

6.7. Let's say that you have a metal bar that is 19.5 cm long and vibrates at A = 440 Hz.

(a) How much would you have to cut off the bar to raise the pitch by a just whole step (9/8)?

(b) Before I cut off the full amount found in (a), I first cut off just 0.75 mm (= 0.075 cm). What is the new frequency and could I hear the change in pitch?

6.8. If you keep going up by perfect fifths, you never reach a perfect octave. This is the basic problem with all methods of tuning. We will look at the minor third, in the same way.

(a) The Pythagorean minor third corresponds to 32/27. How many Pythagorean minor thirds are in an octave? Can you tell the difference between the octave formed by this interval and a perfect octave?

(b) Repeat (a) for a just minor third, 6/5.

(c) Repeat (a) for an equal temper minor third, three half steps.

6.9. (a) Using the Pythagorean system for generating notes on a scale, one could add notes endlessly. However, at some point, you would no longer hear the difference between the notes. How many notes could you just distinguish in a half step? (Or, how many just noticeable differences make up a half step?)

(b) What would be the total number of notes in one octave in this system?

6.10. (a) Suppose you wanted to construct a scale where no two notes ever had overlapping critical bands so that you could play any two notes without producing dissonant intervals. How many notes would be in the scale?

(b) The scale in the previous question works for a xylophone but not for a regular instrument like a violin. Why would any two notes of the "consonant" scale sound consonant with a xylophone but not a violin?

6.11. Here's what would happen if we made a musical instrument out of a 2-D air cavity instead of the normal 1-D cavity.

(a) Fill in the following table for the overtones of a rectangular cavity. You are given the fundamental in each direction. Find the simple overtones and the combination tones.

	$f_1 = 500$ Hz	$f_2 =$
$g_1 = 540$ Hz	$(f_1, g_1) =$	$(f_2, g_1) =$
$g_2 =$	$(f_1, g_2) =$	$(f_2, g_2) =$

(b) Which pairs of frequencies are within a critical bandwidth and what are their beat frequencies?

(c) How would all of these frequencies sound if played together?

6.12. Here's what would happen if we made a musical instrument out of a 2-D air cavity instead of the normal 1-D cavity. The dimensions of the cavity are 30 cm by 40 cm.

    (a) Fill in the following table for the fundamental in each direction and all of the overtones (simple and combination) of the rectangular cavity.

	$f_1 =$	$f_2 =$
$g_1 =$	$(f_1, g_1) =$	$(f_2, g_1) =$
$g_2 =$	$(f_1, g_2) =$	$(f_2, g_2) =$

    (b) Which pairs of frequencies are within a critical bandwidth and what are their beat frequencies?

    (c) How would all of these frequencies sound if played together?

6.13. (a) Start with a pitch of A = 440 Hz. In the Pythagorean scale, a major third is two whole steps. What is the frequency of a Pythagorean major third above A = 440 Hz?

    (b) In an equal temper scale what is the frequency of a major third above A = 440 Hz?

    (c) In the just scale what is the frequency of a major third above A = 440 Hz?

    (d) Can you hear the difference between these pitches?

6.14. Start with a pitch of A = 440 Hz.

    (a) In the Pythagorean scale, what is the frequency of a perfect fifth above A = 440 Hz?

    (b) In an equal temper scale what is the frequency of a perfect fifth above A = 440 Hz?

    (c) In the just scale what is the frequency of a perfect fifth above A = 440 Hz?

    (d) Can you hear the difference between these pitches?

6.15. (a) You are running down a football field at 5 m/s towards a band playing a note at 500 Hz. What note would you hear?

    (b) At 500 Hz, what would be the smallest change in frequency that would be necessary for you to hear a difference (i.e., the just noticeable difference)? Could you hear the change in frequency found in part (a)?

6.16. I read about an effect in speakers which seems peculiar: the claim was that if a speaker is sounding loudly at 100 Hz, it is moving back and forth with a high velocity. At the same time, if the speaker is also sounding a note at 1000 Hz, the higher sound can be Doppler shifted by the motion at the lower frequency, distorting the pitch. We want to estimate this effect.

(a) Assume the speaker is vibrating at 100 Hz with an amplitude of 1 cm. This is a very large amplitude. What would be the maximum speed of the surface of the speaker?

(b) Now consider a tone at 1000 Hz. What would be the Doppler shifted frequency of this note using the speed from (a) and the moving source Doppler shift?

(c) You can detect a frequency change of at most 1/200 of the original frequency. Could you hear the frequency shift from the Doppler effect?

6.17. (a) A very good sprinter can run at 10 m/s. If you run at this speed away from an alarm sounding a frequency of 1200 Hz, what pitch would you hear?

(b) Can you detect this change in pitch? What is the just noticeable difference?

(c) Would the shift in frequency be greater than a half step?

6.18. A police car with a siren at 800 Hz is coming at you at 25 miles/h. However, the sound from the siren also reflects off of a building behind the car and reaches you. You are not moving.

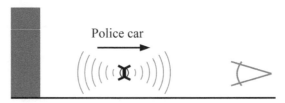

Police car

(a) What frequency do you hear from the siren directly?

(b) What frequency do you hear from the reflected sound?

(c) What is the beat frequency between these two frequencies and how would the pair sound to you?

6.19. Assume that the ear responds in the following way:

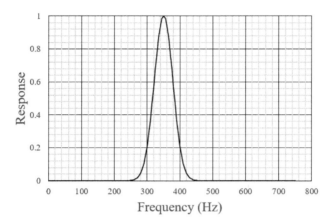

(a) Roughly, what is the center frequency and what is the critical bandwidth?

(b) What is the just noticeable difference (JND) for this center frequency?

(c) Sketch a response curved shifted to a higher frequency by about two half steps. Do the critical bandwidths of the two curves overlap?

6.20. Assume that the ear responds in the following way:

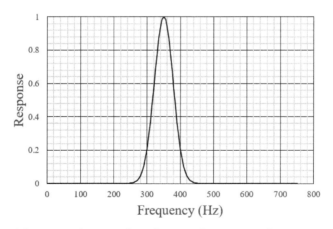

(a) From the graph, what is the center frequency and, roughly, what is the critical bandwidth?

(b) Is this critical bandwidth consistent with our 19% definition? Why or why not?

(c) Sketch a response curved shifted to a frequency lower by two half steps. Do the critical bandwidths of the two curves overlap?

6.21. The critical bandwidth of the ear is 0.190 times the frequency, except that it is never less than 90 Hz.

(a) Start at A = 440 Hz. What would your ear hear if you play this and the note one half step higher?

(b) What would your ear hear if you play this A and the note two half steps higher?

(c) What would your ear hear if you play this A and the note four half steps higher?

(d) Now start at A = 110 Hz. What would your ear hear if you play this and the note one half step higher?

(e) What would your ear hear if you play this lower A and the note two half steps higher?

(f) What would your ear hear if you play this lower A and the note four half steps higher?

6.22. The critical bandwidth of the ear is about 19% of the frequency of the tone heard.

(a) If you start with a frequency of 256 Hz, how much must a second tone be so that the critical bandwidths of the two tones do not overlap?

(b) What is the nearest interval to these two tones?

6.23. You are in a sailboat sailing downwind. Because the boat is traveling with the wind you feel no breeze even though the wind is blowing at 20 miles/h with respect to the ground. You are moving directly away from the shore where your friend is playing a note on a tuba at 120 Hz. (Note: this problem is about sound waves, not water waves!)

(a) With respect to the medium, who is the source, who is the observer, and who is moving?

(b) What frequency of sound does the observer hear?

(c) Could you hear the change in pitch due to the Doppler shift?

Section 6.3

6.24. Two pure tones, when played together, will sound dissonant if:

    A. They are within the just noticeable difference.
    B. The beat frequency is greater than a perfect fifth.
    C. Their critical bandwidths are overlapping.
    D. They form an interval of a diminished sixth.
    E. All of the above.

6.25. The sensation of dissonance is due to what?

    A. The presence of difference tones.
    B. Pain in the ear due to excessive loudness.
    C. Very-high-frequency overtones in a note.
    D. Feedback when two tones are present in the ear.
    E. Neurons trying to fire at two different frequencies.

6.26. Start at 285 Hz and find the note one critical bandwidth higher in pitch. Then find the interval between these two notes. What interval is this closest to:

    A. Half step.
    B. Whole step.
    C. Minor third.
    D. Major third.
    E. Perfect fourth.
    F. Perfect fifth.
    G. Major sixth.

6.27. The critical bandwidth of the ear is 0.190 times the frequency, except that it is never less than 95 Hz.

For each of the following pairs of pure tones, indicate which of these you would hear:

A. A consonant tone and beats; B. One tone and dissonance; C. Two tones and they would sound dissonant; D. Two distinct tones and they would sound consonant.

    (a) A = 440 Hz and the note one (equal temper) half step higher?

(b) A = 440 Hz and the note two (equal temper) half steps higher?

(c) A = 440 Hz and the note four (equal temper) half steps higher?

(d) A = 220 Hz and the note one (equal temper) half step higher?

(e) A = 220 Hz and the note two (equal temper) half steps higher?

(f) A = 220 Hz and the note four (equal temper) half steps higher?

6.28. (a) Arabic music uses scales with many more notes per octave than Western music. Thus, the scale has quarter-steps. This interval corresponds to a ratio of 1.0293. Show that two quarter-steps equals a half step.

(b) If you play a note at 500 Hz could you distinguish this pitch from a note one quarter-step higher?

(c) What would you hear if you played these two notes together?

6.29. (a) Arabic music uses scales with many more notes per octave than Western music. Thus, the scale has quarter-steps. This interval corresponds to a ratio of 1.0293. Show that two quarter-steps equals a half step.

(b) If you play a pure tone at 300 Hz could you distinguish this pitch from a note one quarter-step higher? How would these two notes sound if you played them together?

(c) If you play a pure tone at 800 Hz could you distinguish this pitch from a note one quarter-step higher? How would these two notes sound if you played them together?

6.30. (a) If you go up by a fifth and down by a half step you get a diminished fifth. What is a diminished fifth above 220 Hz?

(b) Find the first three harmonics, including the fundamental, of 220 Hz and the first three harmonics of the diminished fifth.

(c) What pair of overtones (one from each note) are the closest together and how would you perceive these two frequencies?

6.31. (a) Middle A on the piano has a frequency of 440 Hz. The lowest note on the piano is an A four octaves lower. What is the frequency of this note?

(b) What is the frequency of the note one half step higher than the lowest A? What is the beat frequency between these two notes?

(c) How would the ear perceive this beat frequency?

Section 6.4

6.32. When you play the second harmonic on a violin the pitch seems to go up by an octave. The concept of the missing fundamental would seem to say that you should still hear the fundamental as the pitch of the note. Why does the pitch go up and why does not the missing fundamental apply here?

6.33. The interval of a diminished fifth is $\sqrt{2}$ and there are exactly two diminished fifths in an octave.

(a) Show why this is so.

(b) Determine the amount of dissonance between two notes a diminished fifth apart.

6.34. The overtones of a metal bar are at $2.76f$ and $5.40f$.

(a) What are the overtones of a bar with a fundamental frequency of 220 Hz?

(b) What is the fundamental frequency and overtones of a bar tuned on octave higher?

(c) Will these bars sound consonant or dissonant when struck at the same time?

6.35. A peculiar instrument has an overtone series given by: $\sqrt{2^n}$, where f is the fundamental and $n = 0, 1, 2, 3 \ldots$ (Note: $2^0 = 1$).

(a) If the fundamental frequency of a note is 525 Hz, what are the next three overtones?

(b) Now find the pitch and overtones of a note a perfect fifth higher.

(c) If I play these two notes together, how will they sound? Justify your answer: what are the beat frequencies and how would they sound to you?

6.36. (a) Consider a note with a fundamental of 440 Hz (middle A). What is the frequency of the note which is an interval of a perfect fifth higher?

(b) What is the first common harmonic of these two notes?

(c) An augmented fifth is one half step higher than a perfect fifth. What is the frequency of an augmented fifth above 440 Hz?

(d) What is the beat frequency between the same harmonics found in part (b) of the fundamental and the augmented fifth? How would the ear perceive this beat frequency?

6.37. This problem comes from a discussion I had with my piano tuner as he tunes my piano and describes a method for precisely tuning octaves.

(a) Start with A = 220 Hz. What is the frequency of the note (F) a major third lower, according to the just scale?

(b) What is the frequency of the first common overtone of these two notes?

(c) What is the frequency of the note a major third (four half steps) lower than A = 220 Hz, according to the equal temper scale?

(d) What is the beat frequency of between the common overtones found in part (b)? (By common overtone I mean those that are with 10 Hz of each other.)

(e) Finally, what is the first common overtone between the note found in part (c) and A = 440 Hz and what is the beat frequency?

By comparing the results from (d) and (e) you can determine if the A = 220 Hz and the A = 440 Hz form a perfect octave.

6.38. The differences between the equal temper and Pythagorean scale are just on the threshold of human perception. So, they would be hard to tune precisely. However, using beats, these intervals can be tuned precisely, as you will see in this problem.

(a) Start at F = 171 Hz. Find the frequency for the note "A", which is a major third higher, in each tunings:

Pythagorean =

Equal Temper =

(b) What is the just noticeable difference at this frequency? Can you hear the difference between these?

(c) The fifth harmonic of the F and the fourth harmonic of the A are very close. What are the beat frequencies between these harmonics in the two different scales? By listening for this beat frequency you can tune to one scale or the other.

6.39. (a) Start with a pitch of 500 Hz. Find the frequency of the pitch a major third higher according to the just scale. Now calculate all of the harmonics of each note up to 2600 Hz. Compare the frequencies of the two harmonic series and find the closest intervals. Are there any pairs of harmonics that are less than a critical bandwidth (1.19) apart?

(b) Repeat (a) for a major third according to the Pythagorean scale. Which interval would you consider to be more consonant?

6.40. (a) The just major sixth is a factor of 5/3. What is the frequency of a just major sixth above "A" = 440 Hz?

(b) The equal temper major sixth is nine half steps. What is the frequency of an equal temper major sixth above 440 Hz?

(c) Can you distinguish these two frequencies (i.e., what is the just noticeable difference)?

(d) What is the first common harmonic between the "A" and the just major sixth?

(e) Considering just this harmonic, how would the equal temper major sixth sound when played with the "A"?

6.41. (a) Take a note with a fundamental frequency of 420 Hz. What note is an interval of a fourth higher?

    (b) What are the frequencies of the overtones that these notes have in common? (Assume that the overtone series of each note is a complete harmonic series.)

    (c) If the second note is out of tune by being 3 Hz too high, what is the beat frequency between the first common overtones?

6.42. The Pythagorean augmented sixth is formed by going up by perfect fifths 10 times and coming down by octaves. This ends up being a ratio of 59,049/32,768!

    (a) Obviously, some people did not like this definition. The just augmented sixth is formed by a ratio of two integers, both less than 10, that comes the closest to the Pythagorean definition. What ratio defines the just augmented sixth?

    (b) Fill in the following table. On the second line, enter the harmonics of 220 Hz. On the third line, enter the fundamental of a note an augmented sixth higher and enter its harmonics. Don't go past 2000 Hz.

Fundamental	$2f$	$3f$	$4f$	$5f$	$6f$	$7f$	$8f$	$9f$
220 Hz								

    (c) Does the interval of the augmented sixth sound consonant or dissonant? Indicate which pairs of overtones, if any, lie within a critical bandwidth and state how the ear would perceive these pairs.

6.43. Consider the following collections of overtones: 675, 900, 1125, 1350. If these are all played together, what fundamental frequency would you hear? In other words, what is the largest frequency that evenly divides each harmonic?

6.44. Consider the following collections of overtones: 1080, 1800, 2520, ... If these are all played together, what fundamental frequency would you hear? In other words, what is the largest frequency that evenly divides each harmonic?

6.45. The following table gives several examples of which harmonics are present for a variety of (imaginary!) instruments.

In each case, determine what pitch you would hear if the fundamental frequency is $f = 350$ Hz:

	Pitch heard	1f	2f	3f	4f	5f	6f	7f	8f
(a)		X	X		X	X		X	
(b)					X		X		X
(c)			X	X	X		X		X
(d)				X			X		
(e)				X		X		X	

6.46. It is possible that intervals sound more or less consonant depending on which instruments are playing the notes in the interval. Consider an equal temper major third from A440.

(a) Find all of the conflicting harmonics if violins play each of the two notes.

(b) Now assume that a clarinet plays the lower note and a violin plays the upper note. How does this situation compare to part a)?

(c) Now the clarinet plays the upper note and the violin plays the lower note. Is this more or less consonant than the interval in part b)?

(d) Finally, assume clarinets play both notes. Are there any conflicting harmonics?

6.47. A violin has a full harmonic series, while a clarinet has a partial harmonics series ($1f, 3f, 5f, ...$).

(a) For a violin, fill in the following table for the first six harmonics of E = 660 Hz and the harmonics of the note a perfect fifth higher. Which harmonics do these notes have in common?

(b) If the second note were played 10 Hz too low, fill in the last line in the table. Would the interval sound consonant or dissonant?

(c) If a clarinet played the note is part (b) while a violin still played the first note, would it sound consonant or dissonant?

	1*f*	2*f*	3*f*	4*f*	5*f*	6*f*
First note						
Perfect fifth						
Out of tune						

6.48. Start with a note at 525 Hz. Assume that the overtone series of the note is a complete harmonic series.

  (a) At what harmonic do adjacent harmonics lie within a critical bandwidth?

  (b) At what harmonic do adjacent harmonics lie within a just noticeable difference? Can you hear that high a pitch?

Section 6.5

6.49. A major triad consists of a primary note (called the tonic), a second note a just major third higher and a third note a perfect fifth above the primary. Show that all of the difference tones between the fundamentals of each note occur at one or more octaves below the tonic. In other words, all of the difference tones in a major triad reinforce the tonic note.

6.50. If you play a note and a major third above it, what is the difference tone and what interval is it below the original note?

Appendix: Companion Laboratories for *The Music of Physics*

The following labs highlight virtually all of the concepts and equations in the textbook. Many of them can be done at home with readily available parts or with free software. For this reason, the labs can be used in a variety of ways. They can be used in class demonstrations, class group exercises, and homework or to create a full laboratory component to the course.

What is presented here are not complete lab manuals, but rather basic outlines of the labs. They can easily be adapted and expanded, based on the specific needs of the course and instructor. Basic questions are posed to interpret the experiments, but, again, these can be expanded to suit the needs of the particular course. Many of the detailed settings are starting points and can be varied depending on the situation. Finally, many of the labs are fairly short and could be combined in various ways.

An excellent collection of similar acoustic labs can be found in *Phys. Teach.* 54, 500–502 (2016). See bibliography. Lab 5.2 Doppler Shift, in particular, is based on a lab in this collection.

Home Labs (H)

The home labs are physical labs that make use of resources that a typical student has access to. Some easily obtained and inexpensive materials are also used. Many of the home labs have a virtual version, as well. The home labs rely on Audacity and smartphone apps that can display an audio spectrum in real time. Specific apps are mentioned, but there are other choices.

Virtual Labs (V)

The virtual labs make use of an online free software program called Ripple: (http://www.falstad.com/ripple/)

The power of this program is that it is not a "canned" simulator. One must design and draw the experiment in a free-form manner. Thus, each student's experiment may look different and produce somewhat different results. There is also a downloadable standalone version of the app on the website. It runs faster than the web version and may be more convenient.

School Labs (S)

Two of the physical labs require resources which are not easily obtained by individual students.

The vibrating string is a classic lab that is often found in university teaching labs. It is needed to test all aspects of the frequency of a vibrating string.

The rectangular acoustic cavity is a powerful lab that generates a somewhat complicated spectrum, but that is within reach of the students' ability to analyze. It requires some simple machining for a university machine shop. However, there is a very good virtual lab replacement for the physical version.

Chapter 1

1.1 Period vs. Frequency: (H)

Learning objectives:

1) Provide an introduction to Audacity.
2) Understand how formulas in the text can be verified.
3) Show that Eq. 1.1 is valid.
4) Show that period and frequency are not dependent on the shape of the wave.
5) Understand how different methods of measurement can lead to different levels of accuracy. In the following labs, the technique of measuring many periods and dividing by the number of periods will be used to improve the accuracy of a measurement.

Materials:

Computer or laptop with Audacity installed.

Procedure:

1) Open Audacity and set or verify the settings listed below.
2) Create a sine wave as follows: Generate>Tone: Waveform: Sine; Frequency (Hz): 520; Amplitude: 0.8; Duration 0.1
3) Zoom in and select one oscillation and record the period. Does it agree with Eq. 1.1?
4) Now select ten oscillations, record the time, and divide by ten to get the period. Does this give better agreement?
5) One could keep going and count 100 oscillations, but that is hard to keep track of. Instead, select the entire track and choose Analyze>Plot Spectrum then Size: 4096 and Axis: Linear frequency. Finally, mouse over the peak and read the frequency in Peak. Does the agreement get better?
6) Repeat 2) through 5) but with a sawtooth wave instead of a sine wave. Are the results the same? Notice in the spectrum that there are now many peaks. For this lab, only record the lowest frequency peak.

1.2 Loudness vs. Distance (H,V)

Learning objectives:

1) Understand how to test dependences, or how one variable depends on another.
2) Observe how loudness depends on distance in 2D or 3D.
3) Learn how to create and record sounds in Audacity.
4) Learn how to use the Ripple virtual lab simulator.

Physical Lab

Materials:

1) Audacity
2) Earbuds
3) Smartphone with Spectroid app.
4) 12 inch ruler

Procedure:

1) Open Audacity. Set preferences in general Audacity instructions. **Select mono recording**.
2) Click on Generate in the top-level menu.
   Choose Waveform: Sine; Frequency: 2500 Hz; Amplitude: 0.8; Duration: 1 s
3) Play the tone: DO NOT HAVE THE EARBUD IN YOUR EAR! It will be very loud. If you don't hear anything, you need to troubleshoot.
4) In the newly created window, move the L-R slider all the way to R. This sends the tone only to the right earbud and minimizes possible interference from the left earbud. Make sure you know which is the right earbud by listening to each.
5) Download and open the app on your smartphone. The microphone on your phone is on the bottom edge. You need to figure out if it is on the left or the right. Find the position of the earbud that gives the strongest signal.
6) Click on the three vertical dots and then settings. Choose the following settings: Audio>Sampling rate 8000 Hz; FFT size 512 bins; Decimations 5; Desired transform window 20 ms; Exponential smoothing factor 0.65. Display>Frequency axis scale Linear.

7) Place the right earbud and microphone ½ inch apart (Fig. A.1). Start the app and press play in Audacity. Record the maximum signal at 2500 Hz. It should be in the range of –30 dB to –50 dB. While the sound is playing, the app will show the frequency and signal strength of the largest peak. The peak should be at 2500 Hz. Record the signal strength in dB.

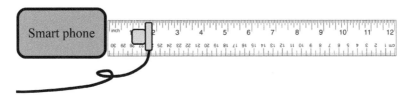

Figure A.1 Physical layout for measuring the dependence of loudness on distance.

8) Repeat these steps for separations between the earbud and the microphone of 1, 2, 4, and 8 inches and record the signal amplitude (dB) in Table A.1—Signal.

Table A.1 Analysis of Loudness vs. Distance in 3D

Distance, D (inch)	Signal, S (dB)	Loudness, L $(10^{S/10})$	$L \cdot D$	$L \cdot D^2$	$L \cdot \sqrt{D}$	L/D
0.5						
1.0						
2.0						
4.0						
8.0						
		Max/Min =				

Analysis:

1) Complete Table A.1. dB is a technical way to measure signal strength, but the details are not important here. The important quantity is Loudness, which is related to the signal, as shown in the table. The other columns represent different possible dependences of Loudness on Distance. You could add your own test cases.

2) The goal is to find some combination of L and D which gives a constant as D is varied. However, we will never find a column that is perfectly constant, as experimental data always has some error. To quantify how constant each column is, we can find the maximum and minimum values. The closer the ratio of the maximum to the minimum values in a column is to one, the more constant the column.

3) Review Table 1.2 and the text following the table to make a conclusion about how Loudness depends on Distance.

Virtual Lab

Materials:

Open the following link, which includes the basic setup:

http://www.falstad.com/ripple/Ripple.html?rol=$+3+512+6
4+0+0+668+0.048828125%0As+2+256+0+0+0.5833325+0
+10+100+1+0%0AP+1+256+258%0A

Procedure:

1) When you use the link above, the program will be set up already. There is a source at the top of the screen and a probe (or microphone) in the middle of the screen. You can interact with the screen, but a more quantitative method is to use the Import/Export option in the File menu. Click on File>Export as text. The line with "s" gives the source parameters and the line with "p" gives the probe parameters. We will leave the source in place and move the microphone. The source is at the top with a vertical coordinate of 0, while the probe is in the middle with a vertical coordinate of 258, to start. The third parameter is the vertical coordinate. There is a window showing the wave and amplitude at the probe.

2) Change the vertical coordinate of the probe to 5 by clicking File>Export as text. Change the third parameter on the "p" line to 5. Then click Re-Import.

3) Record the amplitude at the probe in Table A.2.

4) Repeat 2) and 3) for a vertical coordinate of 10, 20, 40, 80, 160, 320.

Table A.2 Analysis of Loudness vs. Distance in 2D

Distance, D	Amplitude, A	Loudness, L (A^2)	$L \cdot D$	$L \cdot D^2$	$L \cdot \sqrt{D}$	L/D
5						
10						
20						
40						
80						
160						
320						
		Max/Min =				

Analysis:

5) Complete Table A.2. In the simulation, the amplitude of the wave is given. Loudness is equal to the square of the Amplitude, as shown in the table. The other columns represent different possible dependences of Loudness on Distance. You could add your own test cases.

6) The goal is to find some combination of L and D which gives a constant as D is varied. However, we will never find a column that is perfectly constant, as experimental data always has some error. To quantify how constant each column is, we can find the maximum and minimum values. The closer the ratio of the maximum to the minimum values in a column is to one, the more constant the column.

7) Review Table 1.2 and the text following the table to make a conclusion about how Loudness depends on Distance.

8) If you also completed the physical version of this lab, you will notice you get a different dependence. Can you think of any reasons why this is so?

1.3 Tone Quality and Shape of a Wave (H)

Learning objectives:

1) Observe how timbre or tone quality depends on the shape of a wave.
2) Learn how to create sounds in Audacity.
3) Learn how to manipulate tracks in Audacity.

Materials:

    Audacity

Procedure:

1) Open Audacity. Set preferences in general Audacity instructions. **Select mono recording**.
2) Click on Generate in the top-level menu. Choose Waveform: Sine; Frequency: 500 Hz; Amplitude: 0.2; Duration: 2 s. Click out of the track to unselect.
3) Repeat 2) with a square wave and then a saw tooth wave.
4) Play each track individually and then zoom by clicking on the zoom tool. Notice how different each track sounds and how different each wave looks.
5) The square wave and the saw tooth wave are not very pleasant, as they contain very high frequencies. To soften the tone, Click on Effect>Low-Pass Filter. Select Frequency (Hz) 1500 and Roll-off (dB per octave) 36 dB.
6) Now play the square and saw tooth waves after filtering and notice how they are more pleasant. Again, zoom in to see the effect of the filter. The sharp edges have been softened.

1.4 Frequency of a Water Bottle (H)

Learning objectives:

1) Understand how to test dependences, or how one variable depends on another.
2) Determine how the frequency of a water bottle depends on volume. Note: volume, in this context, does not mean loudness. It refers to the empty space in the bottle.

Materials:

1) Clear water bottle with straight sides
2) Smartphone with Spectrum app

Procedure:

1) Mark the water with seven equally spaced lines, starting near the top where the bottle just begins to bend inwards and ending at the lowest point that is still straight. Label the marks 0 through 6, starting at the top. See Fig. A.2.

2) Fill the bottle with water up to the lowest mark (6). Hold the smartphone with the spectrum app running near the bottle. Gently blow across the bottle to make a tone.
3) Record the frequency of the tone in Table A.3.
4) Repeat 2) and 3) filling the bottle to each mark 5, 4, 3, 2, and 1.
5) Enter the results into Table A.3 and complete the rest of the table.
6) State your conclusion as to how frequency depends on volume for a water bottle.

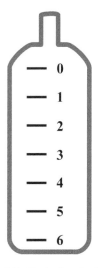

Figure A.2 Marked water bottle.

Table A.3 Analysis of Frequency vs. Volume

Volume, V (mark)	Frequency, F (Hz)	$F \cdot V$	$F \cdot V^2$	$F \cdot \sqrt{V}$	F/V
1					
2					
3					
4					
5					
6					
	Max/Min =				

Chapter 2

2.1 Vibrating String (S)

Learning objectives:

1) Determine how the frequency of a vibrating string depends on length, tension, density, and mode number.
2) Visualize 1D modes, including the placement of nodes and antinodes.

Materials:

1) Flexible but non-elastic string.
2) Weight pan and various weights
3) Pulley
4) Mechanical oscillator
5) Ruler

Procedure:

1) Assemble the apparatus shown in Fig. A.3.

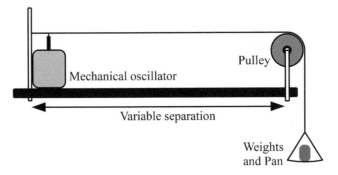

Figure A.3 Apparatus to test a vibrating string.

2) Adjust the frequency of the mechanical oscillator until you see a stable fundamental mode. Record the string length and frequency in Table A.4.
3) Repeat 2) for several string lengths.
4) Complete the rest of Table A.4 and make a conclusion as to how frequency depends on string length, L.

5) Repeat sets 2) – 4), but change the mode number, *n*, for a fixed length, density, and tension. Record the location of the nodes.
6) Repeat sets 2) – 4), but change the tension, *T*, for a fixed length, density, and mode. The tension will be proportional to the weight in the pan.
7) Repeat sets 2) – 4), but change the density, *U*, for a fixed length, tension, and mode. The density can be changed by twisting two, three, or four strings together.

Table A.4 Analysis of frequency vs. string length (note: the same table can be used for tension *T*, density *U*, and mode number *n*)

String length, L (cm)	Frequency, F (Hz)	$F \cdot L$	$F \cdot L^2$	$F \cdot \sqrt{L}$	F/L
	Max/Min =				

2.2 Measure Speed of Sound from Wavelength and Frequency (V)

Learning objectives:

1) Determine the speed of sound from the wavelength and frequency.
2) Verify Eq. 2.1.
3) Become familiar with more tools in the Ripple virtual simulator.

Materials:

Open the following link, which includes the basic set up:

http://www.falstad.com/ripple/Ripple.html?rol=$+3+512+6
4+0+0+662+0.048828125%0As+2+256+0+0+1.7499975+0
+10+100+1+0%0AP+1+331+19%0A

Procedure:

1) Open the Ripple app with the link given above.
2) Let the app run until the screen fills with the waves and then click on Stopped.
3) The frequency slider does not give the true frequency. You can find the actual frequency in the probe window. Enter the frequency in Table A.5.
4) Now measure the wavelength. Click on a dark band of the waves directly below the source. In the lower left hand corner of the app the coordinates of the cursor are displayed. Note the second coordinate, which is the vertical position.
5) Count 10 dark bands going straight down from the source and again note the second coordinate. Take the difference between the two coordinates and divide by 10 to determine the wavelength. Enter this in Table A.5.
6) Repeat 2) – 5) with the frequency slider at 12, 9, 6, and 3. Each time, after setting the slider, deselect Stopped and click on Clear Waves.
7) Finishing filling in Table A.5.
8) Does the average wavelength agree with what you know about sound waves?
9) Is Eq. 2.1 accurate?
10) Why count 10 wavelengths and divide by 10? Why not measure one wavelength?

Table A.5 Analysis of frequency vs. wavelength

Frequency slider	Frequency, F (Hz)	Wavelength (m)	Velocity (m/s)
15			
12			
9			
6			
3			
		Average of v:	

2.3   Closed-Closed tubes (H)

Learning objectives:

1) Determine the resonant frequencies of a closed-closed tube.
2) Explore the validity of Eq. 2.5.
3) Learn how to create a frequency sweep in Audacity and use it to measure modes of an acoustic cavity.

Physical Lab

Materials:

1) Audacity software
2) 30 cm long, 1 or 2 inch diameter PVC tube with a rigid cap on one end.
3) Earbuds
4) Smartphone with spectrum app or earbuds with a microphone

Procedure:

1) Plug in earbuds before opening Audacity
2) Open Audacity with the usual settings and generate a frequency chirp from 1 to 4000 Hz, as follows:
   a. Generate>Chirp: Waveform Sine; Frequency 1 to 4000 Hz; Amplitude 0.8 to 0.8;
   b. Interpolation Linear; Duration 20 s.
3) Choose External mic and headphones
4) *If you are using an earbud with a mic*: Drop the earbud with microphone into the tube and turn the tube over near the end of a table, so that the wire just gets out. See Fig. A.4.
5) Adjust the balance to the left earbud.
6) Start record.
7) Zoom the vertical axis on the recorded track to bring out the peaks. Each peak should correspond to a mode frequency. Record the frequencies.
8) To analyze the frequencies, first estimate the overtone series from the length of the tube. From that, identify the mode number for the measured frequencies.
9) Divide each frequency by its mode number to get the fundamental frequency.

Figure A.4 Closed-closed tube with earbud and microphone inside.

10) The fundamental is often hard to measure directly and is often rather far off from the measures from the higher modes. Average the fundamental derived from the higher modes to get a final measurement of the fundamental. From this, determine the length of the tube and compare to your measured length.

11) *If you are using a smartphone with a spectrum app*: Drop the left earbud into the tube. Close the top with a fairly thin material like a thin piece of wood or plastic.

12) Hold the smartphone vertically against the top of the closed tube with the microphone pointed down. See Fig. A.5.

Figure A.5 Smart phone on tube.

13) Hit play and watch the spectrum carefully to see where the mode frequencies show up. Record each mode that you find. The go on to step 8).
14) Note: the fundamental frequency of this tube will be important for Lab 2.5.

2.4 Register Hole in an Open-Open Tube (V)

Learning objectives:

1) Learn how to find the resonant frequency of an acoustic cavity.
2) Measure the fundamental and second harmonic of an open-open tube.
3) See how the presence of a register hole affects the two modes.

Materials:

Ripple app

Procedure:

1) Open the Ripple app.
2) Click on File>Options. Enter Screen width scale (m): 2
3) Create an open-open tube as follows:
   a. Create a vertical wall about 70 cm long
   b. Create a second vertical wall about 33 cm long, aligned with the top of the first wall, and about 5 cm away.
   c. Create a second vertical wall about 33 cm long, aligned with the bottom of the first wall and right below the second wall.
   d. You should now have an open-open tube with a hole in the middle.
   e. Finally, create a short wall to plug the hole for now.
4) Create a probe by right-clicking on the Ripple canvas and selecting Add probe. Then drag it into the tube about a quarter of the way from the bottom.
5) Create a single source by right-clicking on the Ripple canvas and selecting Add point source. Then drag it slightly above and slightly to the side away from the side with a hole.

6) Estimate the fundamental frequency of the tube from the length and Eq. 2.5. This is going to be too high a frequency, as it does not include the end correction.

7) Right-click on the source and set the frequency to the frequency found in 6). Click on Clear Waves and let the app run.

8) Watch the top number in the Probe window—that is the sound amplitude. It will start to rise as the wave in the tube builds up. It will reach maximum and start to come down. Record the frequency and maximum amplitude.

9) Decrease the frequency by 10 Hz and repeat 7) and 8). As you get closer to the resonant frequency, the buildup twice will get longer and longer. You can also go up in frequency to make sure you have found the peak.

10) Once you found the fundamental, find the second harmonic in the same way. Your starting point will be twice the fundamental. Record enough points to make you found the peak.

11) Once you found the fundamental and second harmonic, reset the frequency to the fundamental and open the hole in the tube. Let the app run and observe what happens to the fundamental mode.

12) Repeat 11) at the second harmonic and observe the effect of the hole on the mode of the second harmonic.

13) Draw some conclusions about the effect of the register hole.

2.5   End Correction for an Open-Closed Tube (H)

Learning objectives:

1) Determine the resonant frequencies of an open-closed tube.
2) Observe the effects of the end correction.

Materials:

1) Audacity software
2) 30 cm long, 1 or preferably 2 inch diameter PVC tube with a rigid cap on one end.
3) Earbuds
4) Smartphone with spectrum app or earbuds with a microphone

Procedure:

1) There are two options for finding the resonant frequencies of the open-closed tube:

    a. Using the method given in Lab 2.3.

    b. Simply hold a smartphone with a spectrogram program running right on top of the open end of the tube. Any background noise will provide acoustic frequencies. The open-closed tube will filter out those frequencies that are resonant with the tube and will show up in the spectrum.

2) Using either method, record as many overtones as you can.

3) For each overtone, you can find a value for the fundamental $(v/(4L))$ using Eq. 2.6. As in Lab 2.3, the higher overtones give a better measurement of the fundamental. Find the average of the fundamentals that you calculated.

4) Again, using Eq. 2.6, find the effective length of the tube from the average value of the fundamental.

5) From Lab 2.3, you can determine the actual length of the tube from the average fundamental in that lab and Eq. 2.5.

6) Finally, the difference between the effect length found for the open-closed tube and the actual length found from the closed-closed tube, the end correction can be determined. How well does it agree with Eq. 2.7?

2.6 Major Triad in Pythagorean, Just, Equal Temper Tuning (H)

Learning objectives:

1) Learn how to construct chords in Audacity.
2) Learn how to choose different tunings.
3) Hear the difference between a major triad in the three main tuning systems.

Materials:

    Audacity

Procedure:

1) Open Audacity

2) Choose a starting note between 300 and 500 Hz. Record in Table A.6.

3) Calculate the frequencies of a note a major third higher in Pythagorean, Just, Equal Temper tuning. Record in Table A.6.

4) Calculate the frequency a perfect fifth higher. Record in Table A.6.

5) Generate a tone for the fundamental, the major thirds, and the perfect.

    a. For each tone, click on Generate>Tone: Waveform Triangle (if you don't see Triangle, update Audacity); Frequency the corresponding frequency in Table A.6; Amplitude 0.4; Duration 2 s.

    b. To soften the tone, click on Effects>Low-pass filter: Frequency 2000 Hz; Roll-off 12 dB.

    c. Make sure that no track is highlighted when you create the next track.

6) To listen to the triad in each tuning, mute two of the three major thirds and make sure the fundamental and perfect fifth are not muted. In this way, go through the three possibilities. Rate each one as to, first, which sounds the worst, and then, of the remaining two, which sounds the best. Enter the rating in Table A.6.

7) This lab is most effective if step 6) is done blind. That is, have a partner choose the different tunings without telling you.

Table A.6 Pitches in a major triad

Starting pitch (Hz):			Rating: Best/worst
Major third (Hz)	Pythagorean:		
	Just:		
	Equal temper:		
Perfect fifth (Hz):			

Chapter 3

3.1   Vibrating Bars (H)

Objectives:

1) Determine how the frequency of a vibrating bar depends on length.
2) Find the location of the nodes of the fundamental mode of vibration.
3) Measure the overtone series of a vibrating bar to show it is not harmonic.

Materials:

1) Smartphone with Spectrogram app.
2) One yard or 1 m ruler.
3) Three metal rods of different length—a 5/16 or 3/8 inch diameter round metal rod works well. Do not use threaded rod. You can find this at any hardware store. Ask for a 36 inch rod cut into three pieces—roughly 16, 12, and 8 inches. These do not need to be exact, as you will measure the actual lengths when you do the lab. They can be steel or aluminum.

Procedure:

Find the locations of the nodes for each bar:

1) Open the Spectrogram app and start it running.
2) Hold one of the shorter bars between your thumb and finger and position the end of the bar near the smartphone microphone. Tap the bar and watch the app for a low frequency to appear. This is the fundamental. Move the point where you hold the bar and try again. You are looking for the point where the low frequency is the loudest and lasts for the longest time. This is the position of the node of the fundamental mode. The pitch will be very low for the longer bar—around 200 Hz.
3) Mark this position and measure the distance from the end. Repeat this for the other two bars.

4) Record the length of each bar in Table A.7, under Length. Record the distance of the node from the end of the bar under Node location.

Find the fundamental and overtone series for each bar:

1) Start the Spectrogram app.
2) Tie a rubber band around the bar, 1/5 of the way from one end and hook the ends of the rubber band around your thumb and forefinger. Place the end of the bar near the microphone.
3) Tap the bar with something wooden (spoon or ruler, for example) or hard rubber. You will see some well-defined peaks pop up in the spectrum.
4) Measure and record the frequencies of each peak you see in the spectrum for each bar. There will always be noise, so just select strong peaks. Enter all of this information into Table A.7. The lowest frequency is the fundamental and the higher frequencies are the overtones. The fundamental should be between 100 and 1000 Hz.

Table A.7 Basic measurements of the vibrating bars

	Length	Node location	Funda-mental	Overtone	Overtone	Overtone	Overtone
Bar #1							
Bar #2							
Bar #3							

Analysis:

1) To analyze this data, we want to remove the unnecessary details. The position of the node should be a constant fraction of the length, so calculate the position of the node as a percent of the length by dividing the distance of the node from the end by the length and multiply by 100 to get percent. Do this for each bar and enter the result into Table A.8.
2) Find the overtone series of each bar by dividing the overtones by the fundamental for each bar. Enter the series into Table A.8.

Table A.8

	Node (% of L)	1st overtone/ fundamental	2nd overtone/ fundamental	3rd overtone/ fundamental	4th overtone/ fundamental
Bar #1					
Bar #2					
Bar #2					

3) As before, to determine how frequency depends on length, we try various arithmetic combinations of frequency and length until we find one that is relatively constant. To do this, fill in Table A.9:

Table A.9

Bar length L (cm)	Frequency F (Hz)	$F \cdot L$	$F \cdot L^2$	$F \cdot \sqrt{L}$	F/L
	Max/Min =				

4) Make a conclusion about how frequency depends on length.
5) Does the overtone series correspond to the expected series?
6) What is the average location of the node of the fundamental?

3.2 Tone Quality—Non-Harmonic Overtones (H)

Learning objectives:

1) Produce tones with non-harmonic overtones in Audacity.
2) Observe the effect on timbre.
3) Observer the effect on normally consonant intervals, like the perfect fifth.

Materials:

  Audacity software

Procedure:

1) Open Audacity
2) Choose a pitch between 400 and 500 Hz. Enter the frequency into Table A.10 along with the next two overtones of a vibrating bar ($2.76f$ and $5.4f$).
3) Calculate the pitch a perfect fifth above the fundamental in 2) and calculate the overtones of the perfect fifth.
4) In Audacity, create three tones which are sine waves with the frequencies from 2) and with amplitudes of 0.4, 0.2, and 0.1, respectively.
5) Select all three tracks and click on Tracks>Mix>Mix and Render to new track.
6) Repeat this for the frequencies in 3).
7) You can now mute or delete the tracks with individual frequencies.
8) Play the new track created in 5). Note that it does not have a clear sense of pitch or a well-blended tone quality.
9) Play the tracks from 5) and 6) together and note that the interval does not sound consonant, even though the perfect fifth is the most consonant interval after the octave.

Table A.10

	Fundamental f (Hz)	1st overtone ($2.76f$)	2nd overtone ($5.4f$)
First note			
Perfect fifth			

3.3 Piano Strings (H)

Learning objectives:

1) Observe how the overtones of a piano string in a grand piano are not perfectly harmonic.
2) Observe how a note one or more octaves higher are tuned to the overtone, not perfect octaves.

Materials:

1) Audacity. Note, Audacity can measure more overtones more accurately than smartphone apps.

2) Access to a grand piano.

Procedure:

1) Open Audacity on a laptop placed on a grand piano. The built-in microphone is sufficient to pick up the sound of the grand piano.
2) In Audacity, click record and play A 440 moderately loudly.
3) In the recorded track, there will be a loud "attack" from when the hammer actually hits the string. Select about 1 s of sound right after the attack.
4) Click Analyze>Plot Spectrum. Choose Axis Linear. Increase size as much as possible.
5) Mouse over the peaks in the spectrum and note the frequency in the Peak box (it finds the peak closest to the cursor).
6) Record the measured fundamental and overtones in the first column of Table A.11.
7) In the second column, calculate the harmonics of the fundamental.
8) Repeat 2) – 6) for an "A" one octave higher, but only record the fundamental in Table A.11.
9) Repeat 8) for the remaining two octaves on the piano.
10) Note how the overtones in the first column get progressively larger than the harmonics in the second column.
11) Note how the octaves follow the overtones, not the harmonics.

Table A.11

Measured	Harmonics	Octaves	

3.4   2D Rectangular Cavities—Virtual Option (V)

Learning objectives:

1) Find the resonant frequencies of a 2D rectangular cavity.
2) Verify the frequencies of combination modes.
3) Learn an advanced method for finding resonant frequencies.

Materials:

    Ripple app

Procedure:

1) Open Ripple app
2) Select Resolution 384 with the slider bar (all the way to the left)
3) Click on File>Options>Screen width scale (m) 2
4) Create a rectangle (Fig. A.6):

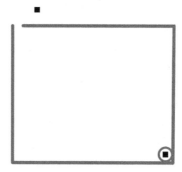

Figure A.6 Example rectangle with source and probe.

  a. Right-click on canvas and select add cavity
  b. Choose the height in the range of 30–50 cm,
  c. Choose the width to be roughly 1.2 times the height. The reason for this comes from Fig. 3.15. This ratio of lengths keeps the various mode frequencies fairly separated.
  d. The cavity needs to be almost closed on top. Right-click on the canvas and select add wall. Close the top except for a small opening in one corner.
  e. Drag the source relatively close to the opening.

    f. Right-click on the canvas and select add probe. Drag the probe into the corner opposite the opening.

5) Based on the width the height of the cavity, calculate the fundamental frequencies in each direction using Eq. 2.5. Enter these in Table A.12 under predicted frequencies. Round all frequencies to integers.

6) The horizontal fundamental will be the lowest frequency. Right-click on the source, select edit and enter this frequency. Round to the nearest 5 Hz.

7) Increase the simulation speed with the slider bar to the maximum value.

8) You can find the actual value of the fundamental using the method in Lab 2.4. Alternatively, you can use the following procedure:

    a. If you are not right on the resonant frequency, the amplitude at the probe will go up and down in time. Stop the app and clear the waves. Right-click on the probe window and select Start recording. Start the app.

    b. Watch the probe to see that the amplitude goes up and down one or two full cycles. When it has, click on stop recording and then click on save recording. Follow the instructions to save the recording.

    c. Open a plotting program like Excel and plot the data in the saved file. It will look something like Fig. A.6. This is called a beat pattern and will be discussed in more detail in Section 4.5. Count the number of cycles that it takes for the wave to build up and come back down. For the example in Fig. A.7, it takes 15 cycles. The fast oscillation is the frequency of the source, in this case 380 Hz. Take the frequency of the source and divide by the number of cycles. In this example, 380 Hz/15 = 25 Hz. This is the difference between the selected frequency and the cavity mode frequency. Since it is just the difference, we don't know whether to add or subtract this value. In the example, the cavity mode frequency will be 355 Hz or 405 Hz. Try adding and subtracting. It will become clear which is the correct one.

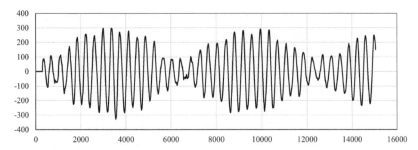

Figure A.7 Beat pattern between driving frequency and cavity frequency.

9) Once you have the measured frequency f enter this into Table A.12 and enter the ratio of measured to predicted frequencies.

Table A.12 Determination of simple modes in each direction showing predicted and measured frequencies and the ratio of measured to predicted

Mode	Predicted	Measured	Ratio
f (Horizontal)			
g (Vertical)			
f_2			
g_2			
f_3			
g_3			

10) Measure the frequency of the vertical fundamental g. This will be off by roughly the same ratio as f, so update the predicted frequency for g by this same ratio. Enter this into the source frequency as see if it is correct.

11) Find f_2. This should be roughly twice the measured fundamental. Enter the predicted value into Table A.12 and find the true value using steps 8) and 9).

12) Find g_2 using step 10).

13) Find f_3 with the predicted frequency being 3/2 of the measured f_2. Similarly for g_3. Enter all values into Table A.12.

14) Enter all the measured values from Table A.12 into the highlighted cells of Table A.13.

15) Calculate the frequencies of the combination modes according to Eq. 3.3.
16) Enter each combination frequency, one by one, into the source. Each time clear the waves and let the mode pattern develop. Verify that the correct mode pattern is created for each combination mode.

Table A.13 Frequencies for a 2D acoustic cavity in Hz

		f (Hz)	f_2	f_3
g (Hz)				
g_2				
g_3				

3.5 2D Rectangular Cavities—School Option (S)

The thin rectangular acoustic cavity presents many interesting and powerful examples of spectroscopy. However, it is slightly out of reach of an individual home lab. The virtual option certainly verifies the frequencies and nodal patterns for all of the modes of a rectangle, but it is not suited to implementing a powerful form of spectroscopy, which will be described here.

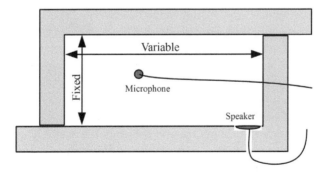

Figure A.8 2D acoustic cavity with speaker and movable microphone.

Figure A.8 shows the acoustic cavity. It is formed with two "L" pieces made out of Plexiglas 6 × 12 inches and 1 inch thick. A flat piece of thinner Plexiglas forms the bottom and top. In one corner of one of the Ls, a speaker is embedded, while a microphone is placed in a hole in the top sheet. With this arrangement, one length of the cavity can be varied smoothly and the microphone position can be moved around to verify the pattern of nodal lines for a particular mode.

The frequencies of the cavity modes can be measured by creating a sound file that sweeps from 0 to 5000 Hz. As this is played, the output of the microphone is recorded. Ideally, the microphone output is rectified and filtered to produce a signal simply proportional to the sound level without the high-frequency component. Such an output is shown in Fig. 6.5. As can be seen, it is a rich spectrum and the assignments of the peaks are not straightforward. Teasing out which peaks correspond to the simple harmonic series in each direction and which are combination modes takes careful consideration. In fact, the data in Fig. 6.5 can be used to determine the dimensions of the rectangular used to take the data and the formula for the combination mode frequencies can be verified. Without any prior understanding, a great deal can be learned by varying the length and making a plot, like in Fig. 3.15. Furthermore, by placing the microphone in highly symmetric locations, like right in the middle of the cavity, certain modes will disappear from the spectrum, greatly narrowing down the possible modes those frequencies correspond to.

This cavity also presents opportunities for studying feedback. With enough gain, certain modes will go into feedback. However, predicting which one is not straightforward, especially as the microphone is moved around the cavity. As the microphone gets moved, the feedback frequency can "hop" between modes.

Once all of the modes are identified and compared to the predicted modes, based on the dimensions of the cavity, the only fitting parameter is the speed of sound. From the measurements, the speed of sound can be determined with sufficient accuracy to be able to measure the temperature of the air in the cavity.

If a light gas, like helium, or a heavy gas, like sulfur hexafluoride, is introduced into the cavity, the mode frequencies will shift accordingly, indicating how a property of the gas can be measured precisely.

Finally, this is obviously a 3D cavity but is analyzed as a 2D cavity. The question is: why does the third dimension not play a role? Clearly, it is because the frequency was not scanned to a high enough frequency. This begs the question: perhaps there is a fourth dimension, but we simply have not scanned to a high enough frequency to detect it?

In summary, this relatively simple acoustic cavity presents many pedagogical opportunities in acoustics, spectroscopy, and precision measurements.

Chapter 4

4.1   Interference in Space—Home Option (H)

Learning objectives:

1) Experience destructive interference.
2) Quantitatively verify Eq. 4.4 for destructive interference.

Materials:

1) Audacity
2) Earbuds
3) Smartphone with Spectrogram app
4) Ruler

Procedure:

1) Attach earbuds to the computer.
2) Open Audacity and click Generate>Tone: Waveform Sine; Frequency 5000 Hz; Amplitude 0.4; Duration 1 s.
3) Select Loop On—the button to the right of Record.
4) Verify that the speaker selection is headphones.
5) Click on play. You should hear the high-frequency 5000 Hz tone through both earbuds. If only one earbud is working, there will be no interference.
6) Hold the earbuds to your right, so you are listening primarily through the right ear.
7) With the earbuds together, you should hear the tone clearly. Move one earbud away from you but in line with the other earbud and your ear. With a separation of 2 to 3 inches, you should hear the tone almost completely disappear. This is a manifestation of destructive interference. If you listen to each earbud individually, you will always hear a tone. However, with both earbuds at the correct separation, the tone will vanish.

8) To measure the conditions for destructive interference quantitatively, place the smartphone on a table and start the spectrogram app.

9) Place a ruler in line with the smartphone microphone and position the two earbuds at the middle of the ruler and the same distance from the smartphone. The earbuds should be about 1 foot from the phone. See Fig. A.9.

10) A frequency of 5000 Hz should appear in the spectrogram on the phone. Sometimes, this arrangement will produce no signal. This means that the earbuds were wired incorrectly and are out-of-phase with each other, already. However, you will still be able to verify Eq. 4.4.

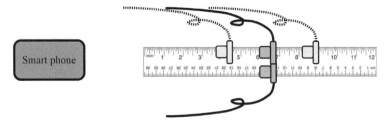

Figure A.9 Setup to observer destructive interference.

11) Move one earbud towards the smartphone until the signal in the spectrogram is a minimum. Note the position of the earbud. Then move the earbud away from the smartphone. Observe the signal go through a maximum and back down to a minimum. Note the new position of the earbud. Find the distance between the points of minimum signal and record that in Table A.14.

Table A.14 Separation between points of destructive interference for different frequencies

Frequency (Hz)	Wavelength (cm)	Distance (cm)
5000		

12) Repeat the experiment for several more frequencies between 3000 and 6000 Hz.
13) Are the points of destructive interference one wavelength apart?

4.2 Interference in Space—Virtual Option (V)

Learning objectives:

1) Determine the conditions necessary for constructive and destructive interferences with the waves going in the same direction.
2) Show that two traveling waves from opposite directions can create interference to produce a standing wave.
3) Produce and interpret interference patterns.

Materials:

Ripple app

Procedure:

First geometry: Same frequency, same direction

1) Open the Ripple App. Click on File>Options Screen width scale (m) 2.
2) Choose Example Two Sources from the top dropdown menu.
3) Choose Color Scheme 2.
4) Set each source to 1715 Hz and move one on top of the other.
5) Calculate the wavelength and enter it into Table A.15.

Table A.15 Source separation for constructive and destructive interference

Wavelength (cm):	Separation of sources (cm)	Ratio of separation to wavelength
First minimum		
First maximum		
Second minimum		
Second maximum		

6) Place a Probe at the bottom center of the screen. Note the strong sound wave everywhere on the canvas.

7) Move one source up until there is destructive interference in the vertical direction. You can either create a dark nodal line or see that the probe signal becomes as small as possible.

8) Record the separation of the sources in Table A.15.

9) Increase the separation of the sources until there is a maximum signal in the vertical direction. Again, record the separation in Table A.15.

10) Repeat 6) – 8) for another point of destructive and constructive interference.

Second geometry: interference patterns

11) As you took the data for the first geometry above, you may have noticed that the pattern of sound waves was getting more complex as you separated the sources. For each separation in Table A.15, look at the overall interference pattern. You will see directions which show constructive interference, much like an antinode in a standing wave and destructive interference, much like nodes. Imagine a circle around the two sources.

12) On the following circles (Fig. A.10), mark the locations of the points of constructive interference with an "x" and the points of destructive interference with a "•" for the four different separations you found above:

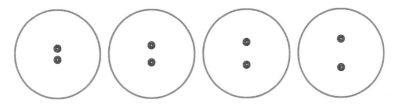

Figure A.10 Points of constructive and destructive interference around two sources.

Third geometry: interference fringes

13) When the separation of the sources gets to be quite large, compared to the wavelength, the pattern gets very complicated. Rather than keep track of every point of constructive and destructive interference, it is more

useful to find the separation of points of constructive and destructive interference. Place the sources off to the left side with a large separation. Measure the separation D. At some distance to the right, L, you will locate a vertical line where you will measure the separation between points of destructive interference. This separation is called y.

14) Record the values from your setup for L, D, and y (λ will be the same as before):

$L =$ _____ $D =$ _____ $y =$ _____

15) The predicted distance y is given by Eq. 4.6. How close is the measured to the predicted value of y?

4.3 Beats (H)

Learning objectives:

1) Verify the phenomenon of beats between two frequencies.
2) Hear how different beat frequencies are perceived by the ear.

Materials:

  Audacity

Procedure:

1) Open Audacity and click on Generate>Tone Waveform Sine; Frequency 400 Hz; Amplitude 0.2; Duration 4 s.
2) In a second track, create the same tone, but with a frequency of 401 Hz.
3) Click on Tracks>Mix>Mix and Render to New Track.
4) This will show a repeating pattern. What is the period and frequency of this pattern? Does it correspond to Eq. 4.7?
5) Play just the mixed track. Can you hear and count the beats?
6) Select the second track and replace it with a frequency of 405 Hz. Again, mix and render the first two tracks.
7) What happens to the beat period and frequency? Can you still hear and count individual beats?
8) Repeat 6), increasing the frequency each time. Find the largest beat frequency where you can still count reliably count the beats.

9) Repeat 8) to find the largest beat frequency where you can still hear individual beats, even if they are too fast to count.

10) Repeat 6) to find the beat frequency that begins to get unpleasant or dissonant.

11) Continue to find how large a beat frequency you need to where the beats are not unpleasant.

12) Do the frequencies you found above conform to the predictions for human perception in Section 4.5?

4.4 Polarization of Light (H)

Learning objectives:

1) Demonstrate that light must be a transverse wave.

2) Show how polarized sunglasses can reduce glare of light from water.

3) Demonstrate a counterintuitive property of waves.

4) Show that computer screens use polarization to control the brightness of the screen.

5) Show how polarizers can reveal mechanical properties of transparent plastic.

Materials:

1) Two pairs of polarized sunglasses or

2) Two pieces of plastic polarizing filters

3) Small transparent plastic box or other material

Procedure:

Crossed polarizers

1) Put on a pair of polarized sunglasses.

2) With the sunglasses on, look through another pair of polarized sunglasses, oriented properly at a well-lit white wall or white piece of paper. Notice the first pair cuts out a fair bit of light, while the second pair does not cut out much more. In other words, with the sunglasses on, the second pair does not change the brightness by much.

3) Rotate the second pair of sunglasses by 90°, so that you are still looking through one of the lenses. Now notice that the scene has become almost completely black—virtually all of the light is gone.

4) This shows light must have two components, or polarizations. The polarizing film absorbs one polarization and lets the other through. If the second polarizer is oriented in the same direction as the first, nothing changes, as the one polarization is already gone. However, when the second polarizer is rotated 90°, it now cuts out the other polarization that got through the first polarizer.

Glare reduction

5) Place a bowl of water on a table and orient yourself so that you can see a light in the reflection of the water. Make sure that the angle at which you are looking at the reflection is a little more than 45°. See Fig. A.11.

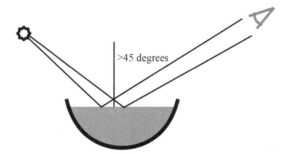

Figure A.11 Arrangement to minimize glare with polarized sunglasses.

6) Put on the sunglasses. The glare may already be gone. Minimize the glare by changing the angle at which you look at the reflection. At some angle, the reflection of the light will almost completely disappear.
7) Take off the sunglasses and look through them rotated by 90°. At this rotation, the glare will hardly be affected.
8) This shows that the glare of water is polarized and can be removed if the polarizer is oriented in the correct direction. When you buy polarized sunglasses, they are always mounted in the orientation that removes glare.

Counter intuitive property of waves

9) Look again at the reflection of the light by the water. Put on the sunglasses and see the reflection disappear, as in 6) above.

10) Place a second pair of sunglasses between you and the water. Rotate the sunglasses from horizontal to vertical. Notice how at 45° the reflection becomes visible again. After the reflection, the light is polarized in the horizontal direction and gets blocked by the sunglasses. However, after the sunglasses at 45°, the light is polarized at 45°. Now, there is a component of vertically polarized light which gets through the sunglasses you are wearing.

Computer screens

11) The fact that light is a transverse wave has many practical applications. Most computer screens use polarization to control the brightness of the screen.

12) Look at a computer screen through a pair of polarized sunglasses oriented correctly. Now rotate your head by 90° while still looking at the screen. Notice how the brightness changes gradually.

Inspection of plastic parts

13) Polarization opens up a new way of looking at things, in particular mechanical stress in clear plastic parts.

14) Wear a pair of polarized sunglasses and look through a clear plastic object at a light. Now place another pair of sunglasses between the light source and the plastic. Rotate the second pair of sunglasses and watch how the colors change.

4.5 Laser Pointers and Diffraction (H)

Learning objectives:

1) Observe how light reflects off of a grating at very shallow angles.

2) Measure the wavelength of the light produced by a laser pointer.

Materials:

3) Laser point, either red or green.

4) Metal ruler with 1/64 inch marking.

Procedure:

5) Arrange the ruler and laser pointer as shown in Fig. A.12.

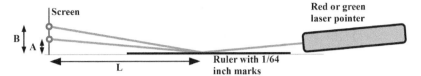

Figure A.12 Arrangement of ruler and laser pointer to show the diffraction of light.

6) The laser beam must strike the ruler at a very shallow angle. The beam should illuminate several inches of the ruler. There will be several spots on the screen. The brightest spot will be the simple reflection off of the ruler. You can tell which one that is by shining the laser on a part of the ruler which has no markings. When you move the laser onto the part edge of the ruler with the 1/64 inch markings, a new spot will appear approximately 1 cm above the reflected spot. There may be one below, as well, but we don't need that one.

7) Record the height of the reflected spot on the screen above the table. That is the value A in the diagram.

8) Record the distance L in cm from where the beam hits the ruler to the screen in Table A.16. Note that since the beam illuminates a fairly long length of the ruler, the measurement needs to be from the center of that length.

9) Record the height of the new beam, B in the diagram. The new beam is the diffracted beam. With much finer divisions, one could rotate the ruler to send this beam straight back to the laser, as in Fig. 4.27. Such components exist and they are called diffraction gratings. However, they are relatively expensive and the diffracted beam seen in this experiment is produced by the same effect. This arrangement just makes the calculations a bit more difficult.

10) d is the spacing of the ruler markings in nm (1 nm = 10^{-9} m). We express it this way because the wavelength of light is best expressed in nm. If the markings are 1/64 inch, then $d = (1/64 \text{ inch})(2.54 \text{ cm/inch})(10^7 \text{ nm/cm}) = 397,000 \text{ nm}$. Enter this in Table A.16.

Table A.16 Measurements for measuring the wavelength of a laser pointer

L (cm)	A (cm)	B (cm)	d (nm)	λ (nm)
			λ actual:	

11) The equation for the wavelength is $\lambda = \frac{d}{2L^2} (B^2 - A^2)$. Using this equation, calculate the wavelength with the values in Table A.16. Green laser pointers have a wavelength of 532 nm and red laser pointers are typically 650 nm. However, the wavelength should be indicated on the laser warning label on the pointer. How well did your measurement agree with the actual value?

Chapter 5

5.1 Gain and Feedback (H)

Learning objectives:

1) Demonstrate feedback in an open-open tube and show that the feedback occurs on a cavity mode.

2) See how the position of the microphone and earbud changes the feedback frequency.

Materials:

1) Open-open tube, roughly 12 inches long and 1 inch in diameter. PVC pipe works well.

2) Pair of earbuds

3) External microphone preferably with a diameter of 1/4 inch or less to fit easily inside the tube. Note, if the computer has a combined 4-pin jack, you will need a 4-pin to microphone/speaker splitter.

Procedure:

1) Open Audacity and select headphones for the speaker and external mic.

2) Click on Generate>Chirp: Waveform Sine; Frequency (Hz) 1 to 5000; Amplitude 0.8 to 0.8; Interpolation Linear; Duration 20 s.

3) In the window with the chirp, pan the balance all the way to the right earbud.
4) Place the right earbud at one entrance to the tube. Place the microphone right at the opposite entrance to the tube.
5) Start recording in Audacity. Make sure that under the Transport menu, Overdub is checked.
6) When the recording is complete, the recording of the microphone output should show a series of peaks.
7) One by one, highlight a short section around each peak. Click Analyze>Plot Spectrum and record the frequency of the peak in Table A.17. This gives the actual mode frequencies of the tube.

Table A.17 Frequencies of an open-closed tube and calculated fundamentals

	Mode 1	Mode 2	Mode 3	Mode 4	Mode 5	Mode 6
Frequency of peak (Hz)						

8) Once the mode frequencies are recorded, you can delete the chirp and recording.
9) Reduce the microphone level to a low value.
10) Go to Settings>System>Sound>Device properties for input>Additional device properties>Listen. Check the Listen to this device box. Select Headphones for Playback through this device. Click Apply, but leave this window open. At this point, feedback is possible.
11) Gradually, turn up the microphone level and speaker level in Audacity until you start hearing feedback. You may experience many different sounds. You are looking for the lowest microphone and speaker setting where you just establish feedback. You are looking for as pure a tone as possible.
12) Once you have a pure tone, press record in Audacity and record the tone for 1 or 2 s. Go back to Microphone properties and uncheck the Listen to this device box. The feedback should stop. Leave all other settings as they are.
13) Select the recording and measure its frequency with Analyze>Plot spectrum. See that the frequency corresponds to one of the modes in Table A.17.

14) Generate a pure tone with a duration of 2 s with the frequency found in 13) and an amplitude of 0.1.

15) Press Record. Note the amplitude of the recorded tone. The ratio of this amplitude to the amplitude of the tone created in 14) is the open loop gain at the frequency where there was feedback. The open loop gain should be very close to 1. This demonstrates that the open loop must be close to 1 for feedback to occur.

16) Repeat 10) through 15), but change the position of the microphone and/or earbud to produce feedback on a different tone. This may require adjustment of the positions and levels of both the microphone and earbud. However, the feedback should jump to a different cavity mode and the open loop gain at that frequency under those conditions should again be close to 1.

5.2   Doppler Shift (H)

Learning objectives:

1) Determine the velocity of a twirling earbud.
2) Record Doppler shifted frequencies.
3) Obtain a precision measurement of the Doppler shifted frequency.
4) Compare the measured and theoretical Doppler shifts.

Materials:

1) A pair of earbuds.
2) A microphone, either built in or external.
3) Audacity.
4) 12" ruler.

Procedure:

1) Open Audacity. Select mono recording.
2) Click on Generate>Tone: Waveform: Sine; Frequency: 5000 Hz; Amplitude: 0.8; Duration: 10 s.
3) Play the tone : DO NOT HAVE THE EARBUD IN YOUR EAR! It will be very loud. If you don't hear anything, you need to troubleshoot.

4) In the newly created window, move the L-R slider all the way to R. This sends the tone only to the right earbud and minimizes possible interference from the left earbud. Make sure you know which is the right earbud by listening to each.

5) Place the earbud about 1 cm from the microphone. Click on Transport in the top-level menu and make sure Overdub has a checkmark (i.e., it is on).

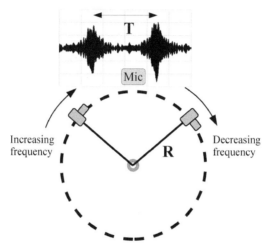

Figure A.13 Arrangement for producing a Doppler shift with sample microphone signal. Regions for measuring the frequency shift are outline in grey.

6) Press record. Adjust the speaker and microphone levels until you get a recorded signal amplitude of at least 0.1. Make sure the signal does not reach 1.0. Measure the exact frequency by selecting part of the signal and clicking on Analyze>Plot Spectrum. Then, either delete that track or click on mute. Otherwise, that track will get played along with the original track. Record this value in Table A.18).

7) Now grasp the earbud cord about 15–20 cm from the earbud and practice twirling the earbud. See Fig. A.13. You don't have to twirl it very fast to get a good result. What you will do is start the recorder in Audacity and start twirling the earbud. Try to get the earbud to pass close to the microphone

without actually making contact. You will see a spike in the amplitude every time the earbud passes close to the microphone. Try it until you get a clean series of about five peaks.

8) Highlight two peaks and measure the time between them, T. See Fig. A.13. Using Table A.18 determine the velocity of the earbud.

9) In Audacity, click on Audio Track to the left of the track you just recorded. Select Spectrogram. Then select Spectrograph Settings from the same menu. Enter the following values: Min frequency: 4890; Max frequency: 5110; Window size: 4096; Gain: 10.

10) Select the region with the maximum upshift in the frequency, using Fig. A.14 as a guide. Measure the upshifted frequency with Analyze>Plot spectrum tool. Enter this under Measured upshifted frequency in Table A.19.

11) Repeat 10) to the right of the peak, light blue lines in Fig. A.14, and enter this under Measured downshifted frequency.

12) Calculate the difference between the upshifted and downshifted frequencies.

13) Calculate the predicted upshifted and downshifted frequencies based on the original frequency and the velocity found in Table A.18. Enter the difference between these.

14) Find the ratio between the measured frequency difference and the predicted frequency difference. Do the measured predicted frequency differences agree with each other?

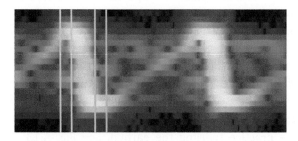

Figure A.14 Spectrogram showing the Doppler Shift. The region between the light green lines has the maximum upshift and the region between the light blue lines has the maximum downshift.

Table A.18 Determination of earbud velocity

Length of earbud cord, R (cm)	
Time between peaks, T (s)	
Velocity of earbud = $2\pi R/T$ (m/s)	

Table A.19 Comparison of predicted and measured Doppler shift

	Measured	Predicted	Ratio
Original frequency (Hz)			
Upshifted frequency (Hz)			
Downshifted frequency (Hz)			
Frequency difference (Hz)			

Chapter 6

6.1 Just Noticeable Difference (JND) and Critical Bandwidth (CB) with Pure Tones (H)

Learning objectives:

1) Measure the JND for your ear.
2) Determine if the JND changes with pitch.
3) Measure the CB for your ear.
4) Determine if the CB changes with pitch.

Materials:

1) Audacity
2) Partner

Procedure:

Just noticeable difference

1) Open Audacity
2) Click on Generate>Tone: Waveform Sine; Frequency (Hz) 500; Amplitude 0.8; Duration 1 s.

3) Scroll the track far enough to the left so that you can place the cursor 0.5 s after the tone.

4) Repeat 2) with 500 (same frequency as a control).

5) On a new track, repeat 2) through 4) with 500 Hz and 501 Hz.

6) Repeat 5) with 500 Hz and 502 Hz.

7) Repeat 5) with 500 Hz and 504 Hz.

8) Repeat 5) with 500 Hz and 506 Hz.

9) You should now have five tracks with two tones each. Have a partner choose randomly from the five tracks and indicate whether you could hear a difference in pitch or not.

10) Compare your responses to the tracks played. What frequency difference was needed to hear the change in pitch? Does this agree with a JND of 0.5%

11) Repeat the entire lab with a frequency of 200, 200.5, 201, 202, and 203 Hz and 1000, 1002, 1004, 1008, and 1012 Hz.

Critical bandwidth

12) Repeat Lab 4.3, but with starting frequencies of 200 Hz, 500 Hz, and 1000 Hz.

13) Interpret the results based on Table 6.1. Table 6.1 is a more accurate prediction of human perception taking into account the critical bandwidth of the ear for different frequency ranges.

6.2 Consonance and Dissonance of Intervals with Complex Tones (H)

Learning objectives:

1) Hear that certain intervals are consonant or dissonant, even when separated by more than a critical bandwidth.

2) Hear the consonant intervals are very sensitive to intonation (how close to the correct frequency a given pitch is played).

3) Hear how consonant intervals depend on the system of tuning.

Materials:

Audacity

Procedure:

Dissonant intervals greater than a critical bandwidth

1) Create complex tone of 500 Hz: Generate>Tone>Waveform Sawtooth; Frequency 500 Hz; Amplitude 0.3; Duration 2 s. To soften the tone, Click on Effect>Low-Pass Filter. Select Frequency (Hz) 1500 and Roll-off (dB per octave) 36 dB.

2) Create a complex tone with a frequency a perfect fifth high.

3) Select both tracks and click Tracks>Mix>Mix and render.

4) Play the track and note that it is still consonant, despite the added harmonics.

5) Repeat 1) – 4), except make the second tone a diminished fifth higher (1.414 in the Equal temper scale).

6) Play the track and note the dissonance, even though the fundamentals are more than a critical bandwidth part.

Dissonance due to intonation:

1) Repeat 1) – 4), but add 10 Hz to the frequency of the perfect fifth.

2) Play the track and note the dissonance, even though a difference of 10 Hz in pure tone intervals would not be perceptible.

Dissonance due to tuning system

1) Repeat 1) – 4), but for a just major third (5/4).

2) Keep the track in 9) and repeat 1) – 4) for a Pythagorean major third (81/64).

3) Here, the difference is somewhat subtle, but alternate playing the tracks to hear that the just interval is cleaner and the Pythagorean is rather unpleasant.

6.3 Missing Fundamental (H)

Learning objective:

> Hear how the ear "fills in" the fundamental frequency based on the overtones.

Materials:

> Audacity

Procedure:

1) Open Audacity

2) Click on Generate>Tone: Waveform Sawtooth; Frequency (Hz) 440; Amplitude 0.8; Duration 1 s. To soften the tone, Click on Effect>Low-Pass Filter. Select Frequency (Hz) 1500 and Roll-off (dB per octave) 36 dB.

3) Scroll the track far enough to the left so that you can place the cursor right after the tone.

4) Repeat 2) with 220 Hz, then repeat 3).

5) Repeat 4) with 110 Hz and then 55 Hz.

6) Select the entire track and press Play. You should hear a tone go down by octaves.

7) Keep this track and create a new track using the same procedure, but with Waveform Sine. Again, use frequencies 440, 220, 110, and 55 Hz.

8) Play each track individually, back and forth a few times. You can mute and unmute each track independently.

9) With the sawtooth wave, you will hear the tone go down by octaves. With the sine wave, you will not hear the last note. This means that you cannot hear 55 Hz, or your system cannot play 55 Hz. It does not matter which. In either case, when you play the sawtooth waves, you hear the last note, but you think it is an octave below the previous note, which means you think you hear 55 Hz. But, you just showed that you cannot hear that frequency or that it is not present. Thus, your brain created the sensation of hearing 55 Hz, as it is the fundamental of the harmonic series that you do hear.

Audacity Notes

The following are settings which will be used for all of the labs based on Audacity.

Edit (PC) Audacity (Mac)>Preferences>

Recording: Play other tracks while recording

Recording: Record on a new track

Tracks>Track Behavior: Advanced vertical zooming

Tracks>Display samples: Connect dots

Bottom of window: Choose "Start and Length of Selection"

Advanced microphone settings: Disable audio enhancements?

Ripple Notes

The virtual lab simulator is at http://www.falstad.com/ripple/. It can be used online, or there is a standalone, offline version. The setup for each lab is given in the link for each lab.

Bibliography

Joshua Allen, Alex Boucher, Dean Meggison, Kate Hruby, and James Vesenka, *Inexpensive Audio Activities: Earbud-based Sound Experiments*, Phys. Teach. 54, 500–502 (2016) https://doi.org/10.1119/1.4965276

Arthur H. Benade, *Fundamental of Musical Acoustics*, 2nd revised ed., Dover Publications, 1990

Donald E. Hall, *Musical Acoustics,* 3rd ed., Brooks/Cole, 2002

Jamie James, *The Music of the Spheres: Music, Science, and the Natural Order of the Universe*, Copernicus, 1993

James Jeans, *Science & Music*, Dover Publications, 1968

Ian Johnston, *Measured Tones: The Interplay of Physics and Music*, 3rd ed., CRC Press, 2009

Robert K. Logan, *The Poetry of Physics and the Physics of Poetry,* World Scientific, 2010

Harry F. Olson, *Music, Physics, and Engineering*, 2nd ed. Dover Publications, 1967

Peter Pesic, *Music and the Making of Modern Science,* MIT Press, 2014

John R. Pierce, *The Science of Musical Sound*, revised ed., Scientific American Books, 1992

Juan G. Roederer, *The Physics and Psychophysics of Music: An Introduction*, 3rd ed., Springer-Verlag, 1995

Colin Ronan, *The Astronomers,* Evans Brothers Limited, 1964

Thomas D. Rossing and Neville H. Fletcher, *Vibration and Sound*, Springer-Verlag, 1995

Charles Taylor, *Exploring Music: The Science and Technology or Tones and Tunes*, IOP Publishing, Ltd, 1992

Index